NIST Special Publication 260-180

Standard Reference Materials®

Certification Report for SRM 2216, 2218, 2219: KLST (Miniaturized) Charpy V-Notch Impact Specimens

Enrico Lucon
Chris McCowan
Ray Santoyo
Jolene Splett
Applied Chemicals and Materials Division
Materials Measurement Laboratory

Jolene Splett
Statistical Engineering Division
Information Technology Laboratory

July 2013

U.S. Department of Commerce
Penny Pritzker, Secretary

National Institute of Standards and Technology
Patrick D. Gallagher, Under Secretary of Commerce for Standards and Technology and Director

Abstract

This Certification Report documents the procedures used to develop certified values of maximum force (F_m) and absorbed energy (KV) for the following KLST (*Kleinstprobe*)-type miniaturized verification Charpy V-notch impact specimens:

- SRM 2216 (low-energy level, $F_m = 2.43$ kN, $KV = 1.59$ J)
- SRM 2218 (high-energy level, $F_m = 1.78$ kN, $KV = 5.64$ J)
- SRM 2219 (super-high energy level, $F_m = 1.79$ kN, $KV = 10.05$ J).

Certified values were obtained from the statistical analysis of an interlaboratory exercise (Round-Robin) conducted among nine international laboratories and coordinated by NIST. The KLST specimens tested in the Round-Robin were machined from previously tested (broken) standard Charpy V-notch verification specimens with energy levels that were low, high and super-high.

This report is intended to provide outside observers with accurate and detailed information on how the materials were certified and how the certification program was conducted. All certified values were established at room temperature (21 °C ± 3 °C).

Full results of the statistical analyses conducted on the Round-Robin data are provided in this Report.

Keywords

Indirect verification; instrumented Charpy testing; KLST; MCVN; miniaturized Charpy specimens; Round-Robin; small specimens; small-scale impact testers.

Table of Contents

1. Introduction

This Certification Report documents the procedures used to develop certified values of maximum force (F_m) and absorbed energy (KV) for miniaturized verification Charpy V-notch impact specimens, at three absorbed energy levels (low, high and super-high).

Miniaturization of mechanical test samples, including impact specimens, is becoming increasingly important as material consumption or material availability often represents limiting factors for mechanical testing.

Historically, miniaturized Charpy V-notch (MCVN) specimens have been used since the 1980s in many countries, mainly as a means of re-using already tested Charpy samples.

The most commonly used miniaturized Charpy specimen is designated KLST (from the German *Kleinstprobe*, or "small specimen"), and has the following nominal dimensions: thickness = 3 mm, width = 4 mm, length = 27 mm, notch depth = 1 mm (i.e., square cross section under the notch); see Figure 1.

The KLST specimen was the first MCVN type to be included in an international test standard, when in 2006 an Amendment (Annex D) titled "*Instrumented Charpy V-notch pendulum impact test of sub-size test pieces*" was approved for inclusion in the ISO 14556:2000 standard. The use of KLST specimens is also allowed by ASTM E2248-13 (*Standard Test Method for Impact Testing of Miniaturized Charpy V-Notch Specimens*).

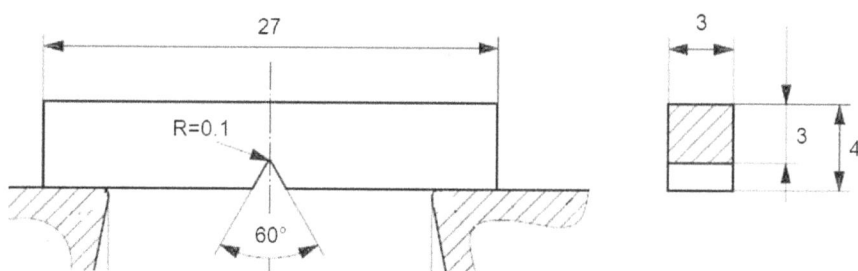

Figure 1 - KLST-type MCVN specimen (dimensions in mm).

KLST specimens can also be tested on a conventional, full-size impact tester provided the anvils and supports are adequately modified to account for specimen dimensions and anvil span. However, the recommended procedure is to use a small-scale pendulum with a significantly lower potential energy (15 J to 50 J instead of 300 J or more) and slightly lower impact speed (3.8 m/s instead of 5 m/s or more). Small-scale impact testers cannot be indirectly verified by means of standard Charpy reference specimens, for both dimensional and energy reasons. On the other hand, MCVN specimens with certified values of absorbed energy or maximum force are unavailable until now. Therefore, users of small-scale impact testers have currently no means of verifying the performance of their machine by the application of an approach equivalent to ASTM E23 or ISO 148-2.

The certified SRMs described in this report are designed to simultaneously verify the performance of the force and energy scales on an instrumented small-scale impact tester at room temperature.

2. Materials and specimens

The KLST-type MCVN specimens used for SRMs 2216, 2218 and 2219 were extracted from previously tested (broken) full-size impact verification specimens of the following lots:

- LL-103 (low-energy level, SRM 2113);
- HH-103 (high-energy level, SRM 2112);
- SH-36 (super-high energy level, SRM 2098).

The certified values of absorbed energy and maximum force for the three lots are shown in Table 1, with expanded uncertainties.

Table 1 - Certified values of absorbed energy and maximum force for the full-size impact specimens from which KLST specimens were extracted.

| SRM | Lot | KV at -40 °C ± 1 °C | | KV at 21 °C ± 1 °C | | F_m at 21 °C ± 1 °C | |
		Certified Ref. Value (J)	Expanded Uncertainty (J)	Certified Ref. Value (J)	Expanded Uncertainty (J)	Certified Ref. Value (kN)	Expanded Uncertainty (kN)
2113	LL-103	15.3	0.1	18.2	0.1	33.00	1.86
2112	HH-103	97.5	0.6	105.3	0.6	24.06	0.70
2098	SH-36	-	-	239.81	1.2	*25.64**	-

Specimens from lots LL-103 and HH-103 were manufactured from AISI 4340 steel bars from a single heat-treated batch in order to minimize compositional and microstructural variations. The chemical composition of the steel is given in Table 2. Additional information on the material and the heat treatment and machining processes can be found in [1].

Table 2 – Chemical composition of AISI 4340 steel, used for lots LL-103 and HH-103 (weight %).

C	Si	Mn	Ni	Cr	Mo	S	P
0.4	0.28	0.66	1.77	0.83	0.28	0.001	0.004

Specimens from lot SH-36 were manufactured from double-vacuum-melted 18Ni-type T-200 maraging steel, adequately forged prior to rolling in order to minimize compositional and microstructural variations. The nominal chemical composition of the steel is shown in Table 3. Additional information on the material and the heat treatment and machining processes can be found in [2].

Table 3 – Nominal chemical composition of type T-200 steel, used for lot SH-36 (weight %).

Ni	Mo	Ti	Al	C	Si	Mn	S	Co	P
18.5	3.0	0.7	0.1	≤0.01	≤0.01	≤0.1	≤0.01	≤0.5	≤0.01

KLST specimens were extracted from broken full-size Charpy samples through the use of EDM (Electro-Discharge Machining), making sure the orientations of the specimen and the notch were preserved (Figure 2). An EDM wire with 0.1 mm (0.004 in) diameter was used to cut the notches.

---

* SH-36 has no certified maximum force value. Therefore, we used as reference value the average of 51 instrumented tests performed at room temperature [3] (F_m = 25.64 kN with standard deviation σ = 0.09 kN).

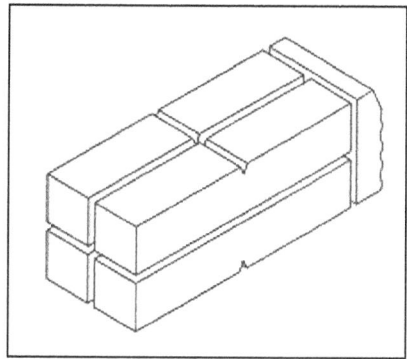

Figure 2 – Extraction of four miniaturized specimens from a broken half of a previously tested, full-size Charpy sample.

3. Description of the Round-Robin

The certified values of absorbed energy and maximum force for the KLST verification specimens were established on the basis of an interlaboratory comparison (Round-Robin). The Round-Robin was conducted among nine international laboratories (Table 4) with well-documented expertise in testing miniaturized impact specimens, and was organized and coordinated by NIST, who provided specimens to all participants.

Table 4 – Round-Robin participants.

Laboratory	Location	Contact person
VTT	Espoo, Finland	M. Valo
KIT	Karlsruhe, Germany	E. Gaganidze
CIEMAT	Madrid, Spain	M. Serrano
SCK•CEN	Mol, Belgium	J.-L. Puzzolante
NRG	Petten, The Netherlands	N. Luzginova
BAM	Berlin, Germany	W. Baer
UJV	Rež, Czech Republic	R. Kopriva
NIST	Boulder, CO, USA	E. Lucon
HZDR	Rossendorf, Germany	H.-W. Viehrig

Each participant received 5 KLST specimens per energy level (low, high and super-high), for a total of 15 specimens.

Tests were to be performed at room temperature (21 °C ± 3 °C) with a small-scale impact tester equipped with an instrumented striker that has a striking edge radius of 2 mm (Figure 3). Tests were conducted in accordance with[†] ISO 14556:2000, ASTM E2248-09 and ASTM E2298-09.

[†] The operational procedures of ISO 14556, ASTM E2248, and ASTM E2298 are practically identical. The most notable differences are in some of the symbols used (for example, the force at unstable crack propagation is designated F_{bf} in the ASTM standards, and F_{iu} in the ISO standard).

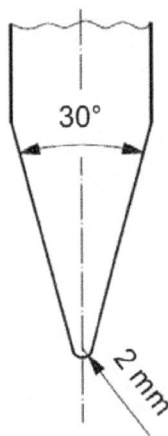

Figure 3 – 2-mm striker used for performing impact tests on KLST specimens.

A standardized fillable PDF (Portable Document Format) form was provided to the participants, so that basic information about their equipment and testing procedure could be collected, along with the test results. A summary of the information provided is presented in Table 5.

Table 5 – Test information provided by Round-Robin participants.

Laboratory	Machine design	Machine capacity (J)	Impact velocity (m/s)	Striker radius (mm)	Anvil span[‡] (mm)	Test T (°C)
BAM	SSP	50	3.85	2	22	21
NIST	SSP	50.15	3.52	2	22	21.4
VTT	SSP	51	3.85	2	22	23
KIT	SSP	24.96	3.85	2	22	21
CIEMAT	SSP	25.01	3.85	2	22	21.5
SCK•CEN	SSP	15	3.71	N/R	N/R	22
NRG	SSP	50	3.85	2	22	21
UJV	N/R	N/R	N/R	N/R	N/R	N/R
HZDR	SSP	15	3.80	2	22	22

NOTE: SSP: small-scale pendulum; N/R = not reported.

Note that no information was provided by UJV, and some data were missing for SCK•CEN[§]. The general observations listed below pertain to participants that provided information.

- All participants used a small-scale impact tester.
- Test machines of three capacity levels were used, namely 15 J, 25 J and 50 J.
- All participants used a 2-mm instrumented striker.
- All participants reported an anvil span of 22 mm.
- Test temperatures were in the range between 21 °C and 23 °C.

    For each test performed, labs were requested to report the following data:

[‡] Nominal value for the KLST anvil span is 22 mm.
[§] These two labs did not return the standardized PDF form.

- <u>characteristic force values</u>: force at general yield F_{gy} and maximum force F_m; in addition (only at the low-energy level), force at unstable crack propagation[**] F_{bf} and crack arrest force F_a;
- <u>characteristic absorbed energy values</u>: energy at general yield W_{gy}, at maximum force W_m, and at test termination W_t; in addition (only at the low-energy level), energy at unstable crack propagation[††] W_{bf} and at crack arrest W_a;
- absorbed energy value (*KV*) provided by the machine dial and/or encoder.

Optionally, participants were also asked to measure and report lateral expansion (LE) and shear fracture appearance (SFA). Finally, labs were required to provide force-time and/or force-displacement curves for all tests, in ASCII or Excel format.

4.   Analyses of the Round-Robin results

4.1   Introduction

For the statistical analyses of the Round-Robin results, only the following parameters were considered: F_{gy}, F_m, W_t, and *KV*. Moreover, certified values were derived for only maximum force (F_m) and absorbed energy (*KV*). For complete results provided by each Round-Robin participant, refer to Annex A.

   In the analyses, participating labs were identified with numbers from 1 to 9 to guarantee anonymity. Note that the numerical sequence does <u>not</u> correspond to the lab order in Table 4 (*i.e.*, the first lab in Table 4 does not correspond to participant #1, and so on).

   The statistical analyses were conducted primarily in accordance with ISO 5725-2:1994 (*Accuracy (trueness and precision) of measurement methods and results*) and ISO/TR 22971:2005 (*Practical guidance for the use of ISO 5725-2:1994 in designing, implementing and statistically analysing interlaboratory repeatability and reproducibility results*). However, some analytical steps also reflect the provisions of ASTM E691-12 (*Standard Practice for Conducting an Interlaboratory Study to Determine the Precision of a Test Method*).

4.2   Presentation of the raw data

The test results supplied by the participating labs are collected in Table 6 (all raw data), Table 7 (mean values), and Table 8 (spread of the values, *i.e.,* standard deviations). Note that Lab #9 did not report F_{gy} values for low-energy specimens.

   The results are plotted versus lab number (raw data plots) in Figure 4 (low energy), Figure 5 (high energy), and Figure 6 (super-high energy). This type of representation provides an instant "picture" of the results, as well as an appraisal of deviations between individual labs or unusual data spread.

   Another common type of graphical data display is the "box-and-whisker" plot, where medians, quartiles, and outliers for each participant are clearly illustrated; see Figures 7-9. In a typical "box-and-whisker" plot, outliers are defined as values larger than $Q_3 + 1.5 \cdot IQR$ or smaller than $Q_1 - 1.5 \cdot IQR$, where Q_1 and Q_3 are the first and third quartiles of the data and IQR, the inter-quartile range, is $Q_3 - Q_1$.

[**] This parameter is designated F_{iu} in ISO 14556:2000.
[††] This parameter is designated W_{iu} in ISO 14556:2000.

Table 6 - Collation of all raw data.

	Low Energy					High Energy					Super-High Energy			
Laboratory	F_{gy} (kN)	F_m (kN)	W_t (J)	KV (J)	Laboratory	F_{gy} (kN)	F_m (kN)	W_t (J)	KV (J)	Laboratory	F_{gy} (kN)	F_m (kN)	W_t (J)	KV (J)
1	1.41	2.32	1.32	1.55	1	1.39	1.75	4.85	5.53	1	1.14	1.77	8.83	10.15
	1.58	2.35	1.36	1.57		1.37	1.74	4.85	5.57		1.18	1.74	8.91	10.16
	1.58	2.34	1.31	1.51		1.40	1.76	4.92	5.64		1.21	1.77	9.17	10.44
	1.48	2.34	1.33	1.52		1.35	1.72	4.81	5.53		1.19	1.79	8.95	10.29
	1.52	2.32	1.28	1.46		1.36	1.72	4.95	5.65		1.14	1.76	9.03	10.26
2	1.95	2.37	1.48	1.60	2	1.50	1.83	5.84	5.88	2	1.40	1.78	9.88	9.97
	1.99	2.35	1.37	1.49		1.63	1.78	5.87	5.86		1.52	1.78	9.99	10.02
	1.95	2.37	1.41	1.51		1.40	1.77	5.46	5.57		1.41	1.81	9.48	9.48
	1.57	2.38	1.46	1.58		1.46	1.80	5.43	5.47		1.42	1.79	10.18	10.21
	1.87	2.31	1.44	1.55		1.49	1.80	5.62	5.72		1.49	1.79	10.29	10.25
3	2.09	2.26	1.55	1.84	3	1.36	1.70	5.31	5.73	3	1.51	1.68	9.85	10.00
	2.14	2.17	1.37	1.72		1.37	1.67	5.41	5.66		1.55	1.70	9.69	10.17
	2.16	2.17	1.33	1.72		1.43	1.67	5.33	5.67		1.52	1.69	9.98	10.42
	2.26	2.26	1.43	1.82		1.38	1.64	5.51	5.82		1.49	1.69	9.72	10.17
	2.04	2.29	1.45	1.79		1.40	1.67	5.51	5.86		1.54	1.70	9.79	10.00
4	2.01	2.83	1.57	1.62	4	1.63	1.90	5.93	5.79	4	1.71	1.93	10.59	10.13
	1.94	2.79	1.44	1.50		1.63	1.92	5.76	5.62		1.69	1.90	10.64	10.05
	1.95	2.75	1.50	1.58		1.56	1.89	5.64	5.43		1.69	1.89	10.14	9.60
	1.96	2.80	1.54	1.58		1.61	1.93	5.89	5.75		1.70	1.91	10.29	9.73
	2.03	2.85	1.56	1.59		1.61	1.88	5.79	5.58		1.69	1.90	10.32	9.79
5	1.86	2.69	1.44	1.35	5	1.57	1.91	6.43	5.52	5	1.67	1.95	11.53	9.96
	1.93	2.72	1.45	1.39		1.68	1.98	6.27	5.38		1.67	1.97	10.77	9.62
	1.93	2.69	1.57	1.52		1.63	1.89	6.30	5.45		1.65	1.98	11.29	10.07
	1.74	2.73	1.50			1.48	1.89	6.20	5.40		1.66	1.97	10.91	9.87
	2.18	2.70	1.43	1.34		1.68	1.98	6.45	5.57		1.65	1.97	11.06	9.87
6	1.58	2.27	1.45	1.80	6	1.32	1.70	5.49	5.90	6	1.41	1.71	9.59	10.00
	1.54	2.24	1.43	2.00		1.36	1.71	5.43	5.80		1.42	1.71	9.76	10.10
	1.55	2.26	1.64	2.00		1.37	1.70	5.57	6.00		1.43	1.70	10.25	10.70
	1.44	2.25	1.48	1.90		1.39	1.71	5.48	5.90		1.42	1.68	10.05	10.50
	1.41	2.28	1.43	1.80		1.36	1.70	5.58	6.00		1.42	1.69	10.09	10.50
7	1.77	2.34	1.46	1.55	7	1.39	1.67	5.74	5.53	7	1.45	1.64	9.42	9.62
	1.67	2.31	1.46	1.53		1.37	1.66	5.34	5.43		1.42	1.69	9.90	10.01
	1.80	2.40	1.37	1.42		1.31	1.69	5.37	5.51		1.44	1.68	9.93	10.01
	1.77	2.35	1.37	1.43		1.33	1.65	5.43	5.54		1.43	1.69	9.43	9.60
	1.79	2.33	1.42	1.49		1.37	1.66	5.33	5.46		1.42	1.66	9.92	9.86
8	2.04	2.45	1.67	1.63	8	1.25	1.80	5.92	5.78	8	1.21	1.79	10.03	9.87
	2.00	2.46	1.65	1.61		1.38	1.77	5.74	5.61		1.26	1.79	9.98	9.86
	1.72	2.36	1.58	1.55		1.52	1.78	5.71	5.59		1.41	1.80	10.47	10.28
	2.04	2.35	1.58	1.54		1.25	1.74	5.58	5.46		1.28	1.79	10.19	10.06
	2.01	2.47	1.72	1.68		1.32	1.75	5.76	5.64		1.31	1.79	10.40	10.27
9		2.43	1.53	1.56	9	1.36	1.85	5.99	5.81	9	1.44	1.84	10.43	10.11
		2.45	1.47	1.52		1.37	1.84	5.67	5.57		1.45	1.84	10.62	10.26
		2.39	1.40	1.43		1.38	1.84	5.75	5.59		1.40	1.85	10.12	9.86
		2.44	1.53	1.54		1.37	1.83	5.74	5.62		1.45	1.86	10.20	9.88
		2.43	1.51	1.54		1.40	1.84	5.68	5.49		1.44	1.85	10.82	10.46

Table 7 - Collation of the mean values for the data in Table 6.

	Low Energy					High Energy					Super-High Energy			
Laboratory	F_{gy} (kN)	F_m (kN)	W_t (J)	KV (J)	Laboratory	F_{gy} (kN)	F_m (kN)	W_t (J)	KV (J)	Laboratory	F_{gy} (kN)	F_m (kN)	W_t (J)	KV (J)
1	1.51	2.33	1.32	1.52	1	1.37	1.74	4.88	5.58	1	1.17	1.77	8.98	10.26
2	1.87	2.35	1.43	1.55	2	1.50	1.79	5.64	5.70	2	1.45	1.79	9.96	9.99
3	2.14	2.23	1.43	1.78	3	1.39	1.67	5.41	5.75	3	1.52	1.69	9.81	10.15
4	1.98	2.80	1.52	1.57	4	1.61	1.90	5 80	5.63	4	1.70	1.90	10.40	9.86
5	1.93	2.71	1.48	1.40	5	1.61	1.93	6.33	5.46	5	1.66	1.97	11.11	9.83
6	1.50	2.26	1.49	1.90	6	1.36	1.70	5.51	5.92	6	1.42	1.70	9.95	10.36
7	1.76	2.35	1.42	1.48	7	1.35	1.67	5.44	5.49	7	1.43	1.67	9.72	9.82
8	1.96	2.42	1.64	1.60	8	1.34	1.77	5.74	5.61	8	1.29	1.79	10.22	10.07
9		2.43	1.49	1.52	9	1.36	1.84	5.77	5.62	9	1.34	1.85	10.44	10.11

Table 8 - Collation of the spread values (standard deviations) for the data in Table 6. Maximum spread values are highlighted in red.

	Low Energy					High Energy					Super-High Energy			
Laboratory	F_{gy} (kN)	F_m (kN)	W_t (J)	KV (J)	Laboratory	F_{gy} (kN)	F_m (kN)	W_t (J)	KV (J)	Laboratory	F_{gy} (kN)	F_m (kN)	W_t (J)	KV (J)
1	0.072	0.013	0.029	0.042	1	0.021	0.018	0.057	0.058	1	0.031	0.018	0.129	0.118
2	0.171	0.028	0.043	0.046	2	0.087	0.025	0.208	0.179	2	0.054	0.012	0.315	0.307
3	0.083	0.056	0.084	0.056	3	0.028	0.021	0.095	0.089	3	0.022	0.009	0.115	0.172
4	0.038	0.039	0.053	0.044	4	0.029	0.022	0.114	0.144	4	0.009	0.014	0.212	0.223
5	0.161	0.018	0.058	0.083	5	0.085	0.046	0.107	0.080	5	0.010	0.011	0.303	0.195
6	0.074	0.016	0.088	0.100	6	0.025	0.005	0.064	0.084	6	0.007	0.013	0.267	0.297
7	0.051	0.036	0.047	0.058	7	0.033	0.015	0.172	0.045	7	0.014	0.020	0.270	0.199
8	0.135	0.057	0.060	0.059	8	0.110	0.025	0.121	0.115	8	0.074	0.007	0.219	0.207
9		0.023	0.055	0.051	9	0.097	0.007	0.130	0.119	9	0.081	0.008	0.290	0.255

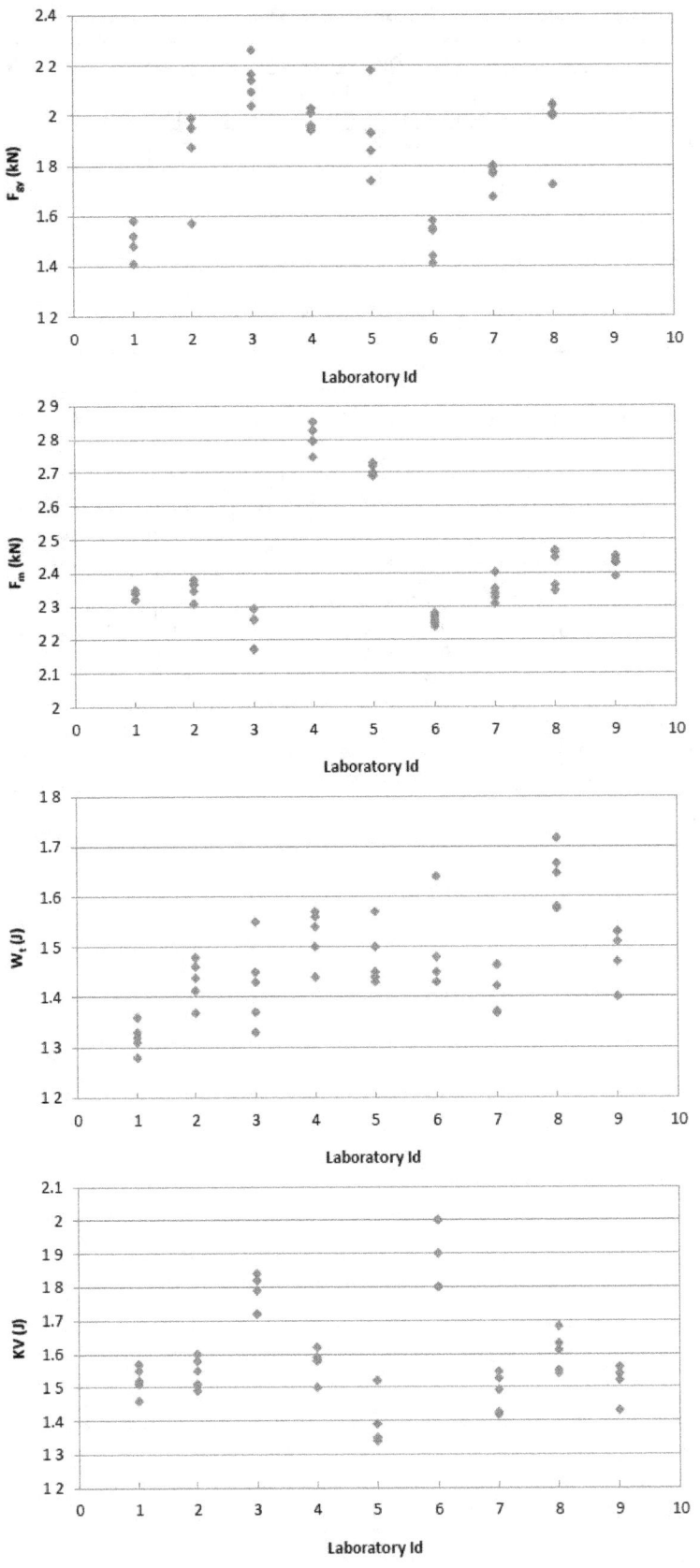

Figure 4 - Graphical representation of the raw data for the low-energy specimens.

8

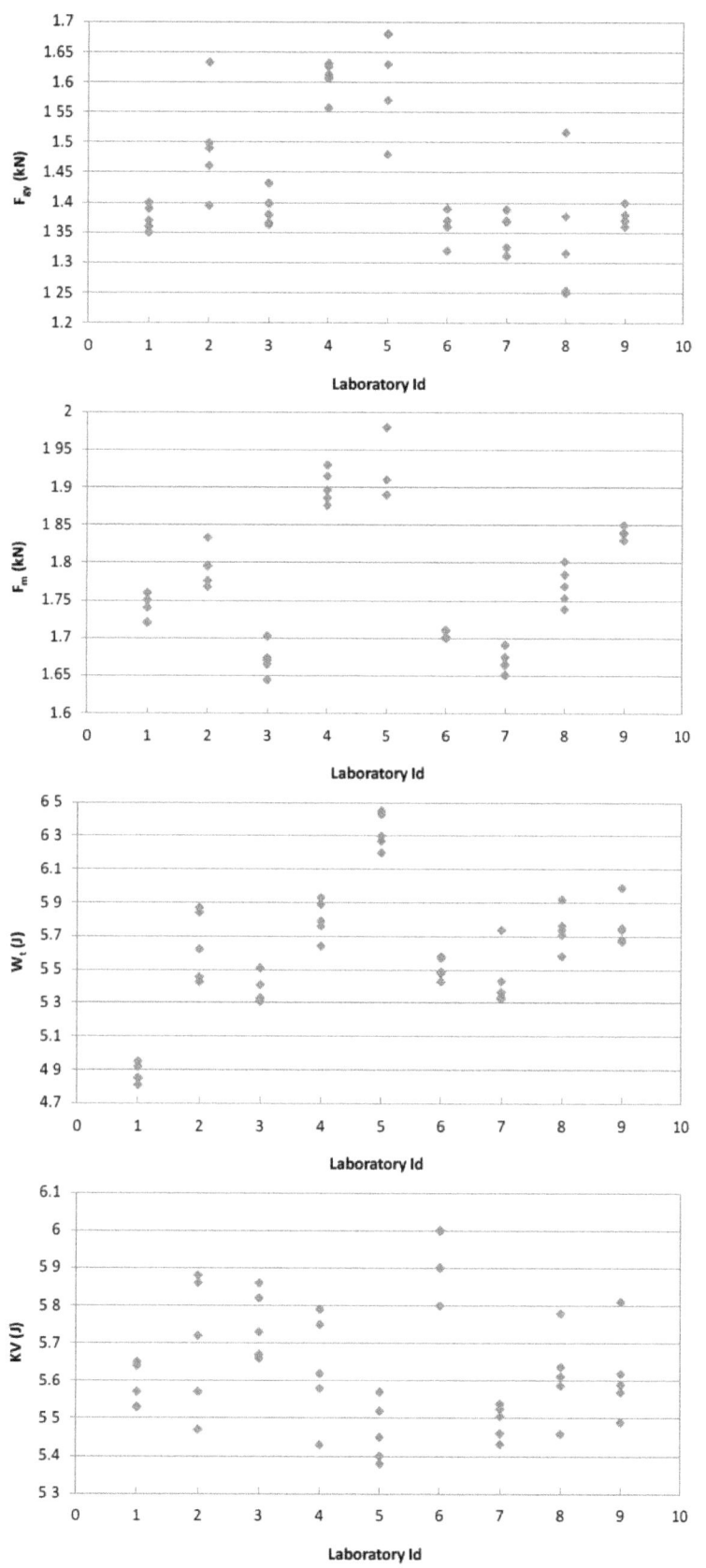

Figure 5 - Graphical representation of the raw data for the high-energy specimens.

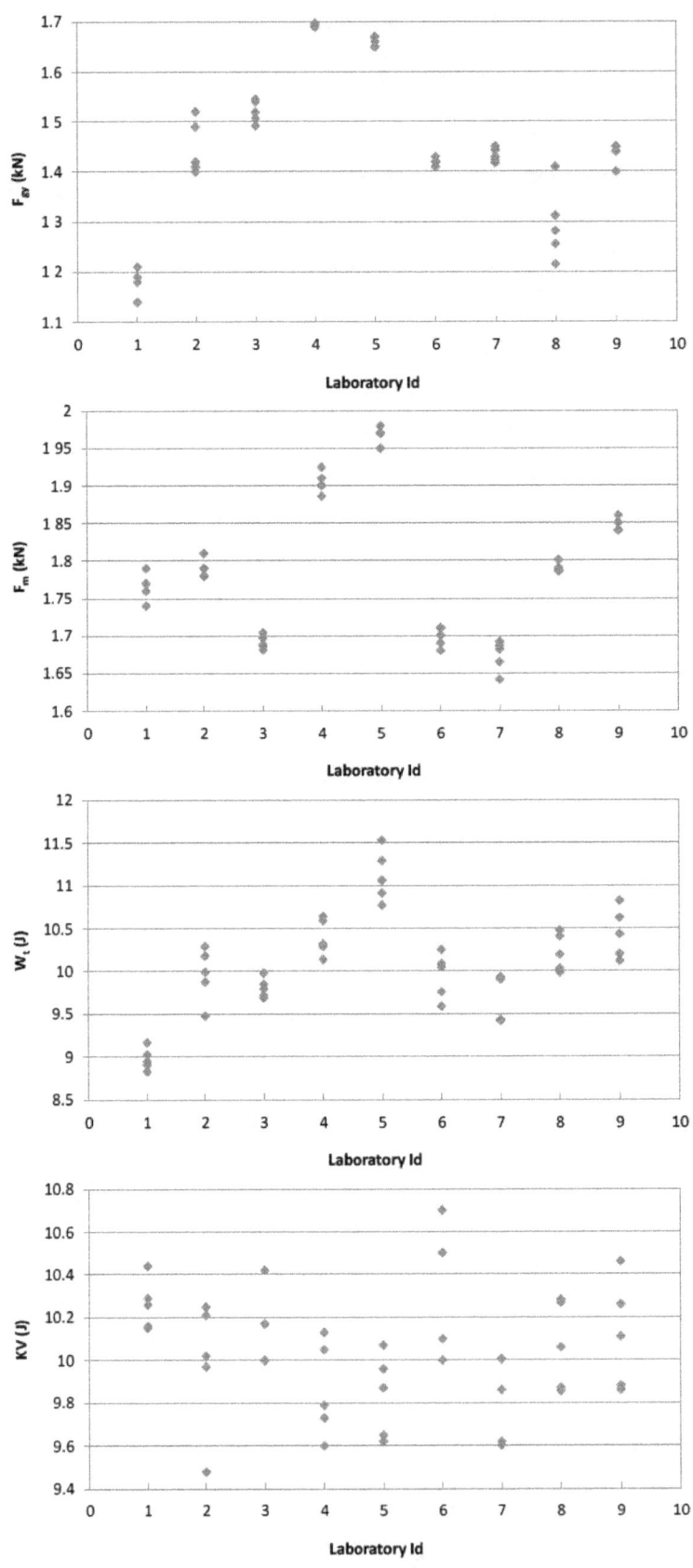

Figure 6 - Graphical representation of the raw data for the super-high energy specimens.

10

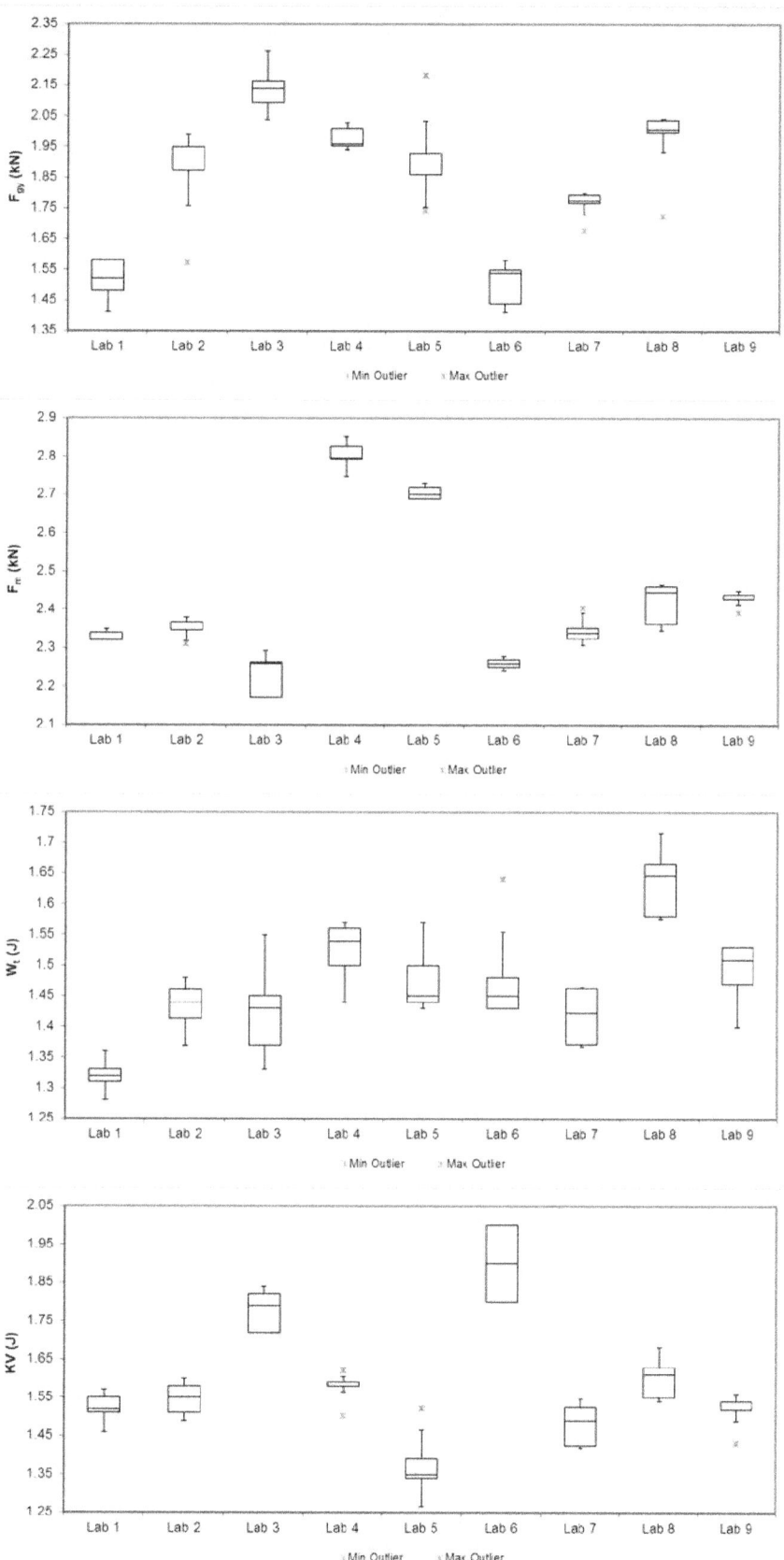

Figure 7 – "Box-and-whiskers" plots for the low-energy specimens.

11

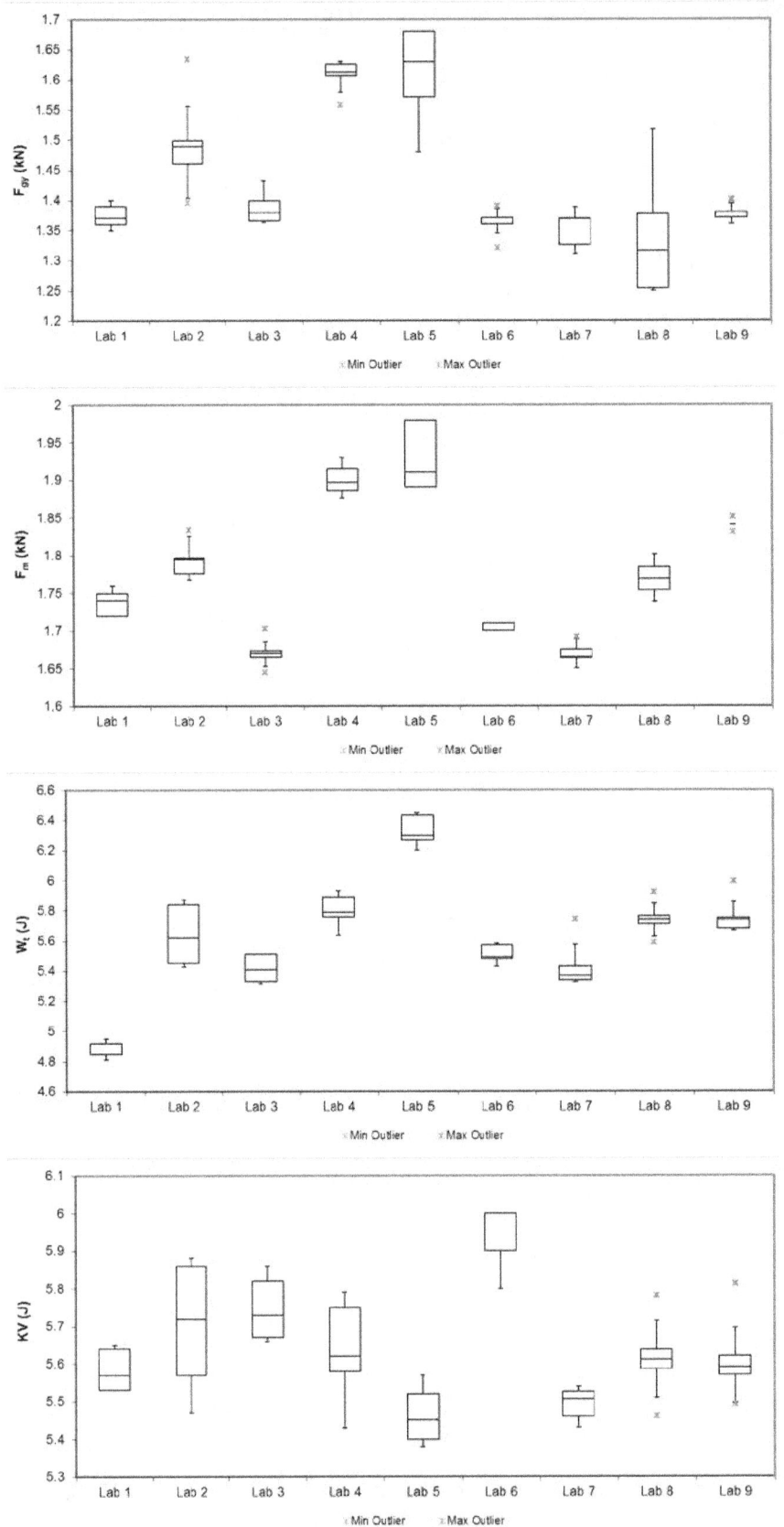

Figure 8 – "Box-and-whiskers" plots for the high-energy specimens.

12

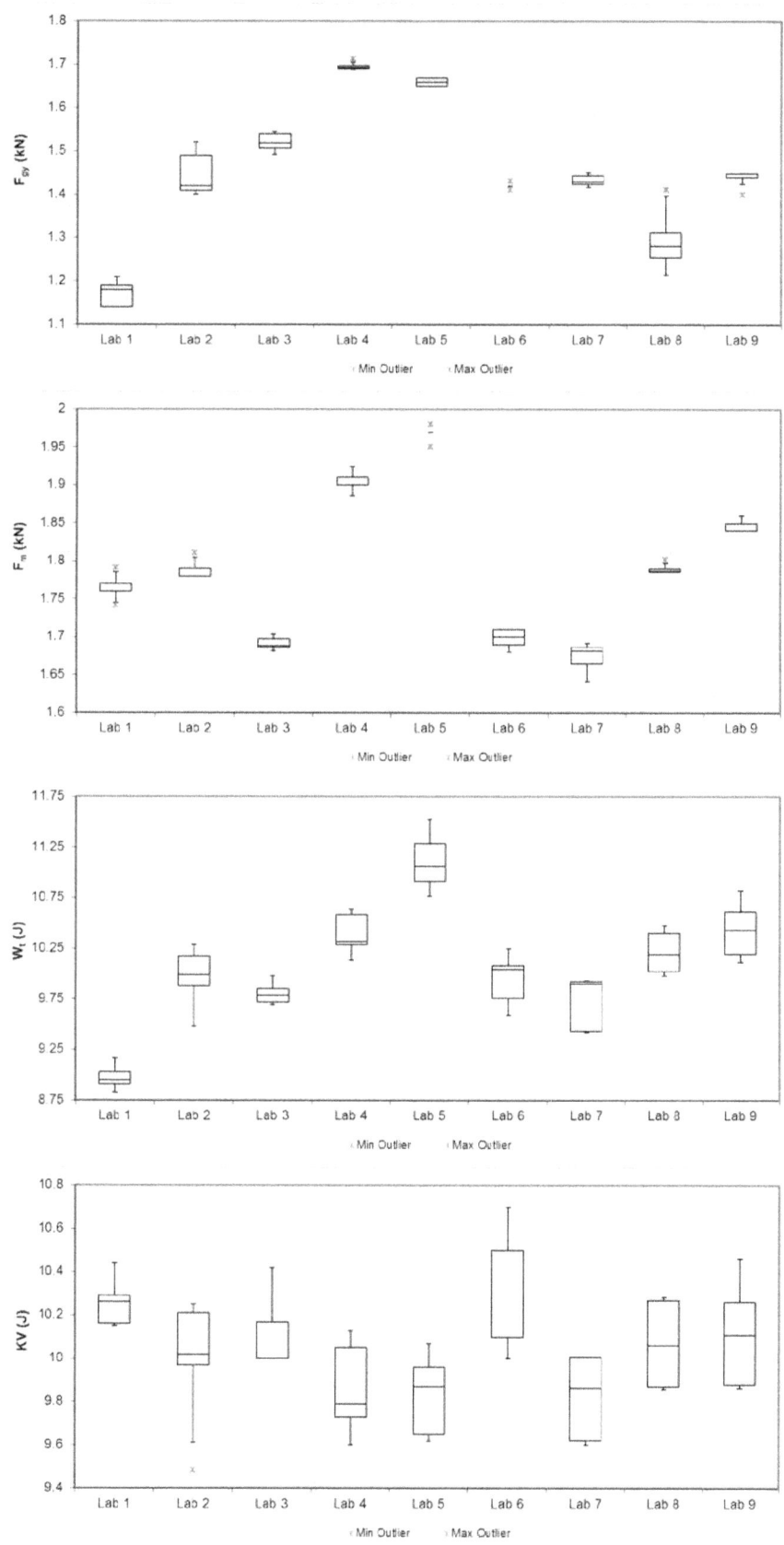

Figure 9 – "Box-and-whiskers" plots for the super-high energy specimens.

4.3   Mandel's plots of *h* and *k* statistics

Repeatability and reproducibility standard deviations are established from data collected for each of the investigated parameters. However, the presence of individual laboratories or values that appear to be inconsistent with the general trend may affect the calculations, and decisions need to be made with respect to these suspected outlier values.

ISO 5725-2:1994 introduces two measures, originally proposed by Mandel [4]:

- the between-laboratory consistency statistic, *h* (ratio of the difference between the lab mean and the mean of all labs, and the standard deviation of the means from all labs);
- the within-laboratory consistency statistic, *k* (quotient of the lab standard deviation and the mean standard deviation for all labs).

Both statistics can be used to describe the variability of the testing method.

For the individual energy levels, Mandel's *h* and *k* plots are given in Figures 10 through 12 and Figures 13 through 15, respectively. Based on a graphical inspection of the plots, one can identify individual results for each laboratory that might be considered different from the expected distribution of results. To facilitate this, dotted and dashed lines are also plotted, corresponding to critical values of Mandel's statistics at the 95 % and 99 % confidence levels.

From the examination of Figures 10 through 15, the following emerges:

- At the low-energy level (Figures 10 and 13):
  - none of the participants exceeds the 99 % critical value of *h* or *k* for any of the parameters;
  - Lab #4 (F_m), Lab #8 (W_t), and Lab #6 (KV) show significant positive deviations (high *h*) with respect to the overall means at the 95 % confidence level;
  - Lab #2 (F_{gy}), Labs #3 and #8 (F_m), and Lab #6 (KV) show significant spread (high *k*) at the 95 % confidence level.
- At the high-energy level (Figures 11 and 14):
  - Lab #6 shows a significant positive deviation (high *h*) for KV at the 95 % confidence level, but does not exceed the 99 % confidence level;
  - the values of *k* for Lab #8 (F_{gy}) and Lab #2 (W_t and KV) are higher than the 95 % critical value but lower than the 99 % critical value;
  - Lab #5 shows a very significant spread for F_m (*k* value higher than the 99 % critical value).
- At the super-high energy level (Figures 12 and 15):
  - none of the participating labs shows significant deviations from the mean trend for any of the parameters (all *h* values are below the 95 % confidence level);
  - Lab #8 (F_{gy}) and Lab #7 (F_m) yield *k* values above the 95 % critical value but below the 99 % critical value;
  - Lab #9 shows very significant spread for F_{gy}, with *k* higher than the 99 % critical value.

As a general remark, Mandel's statistics do not indicate specific laboratories exhibiting patterns of results that are markedly different from everyone else's, as would be indicated by consistently high or low within-laboratory or between-laboratory variations across all measured parameters. Nevertheless, the presence of a few extreme values of *h* and *k* (mostly at the 95 % confidence level) warrants the search for possible outliers through the use of numerical outlier techniques, as detailed in the following section.

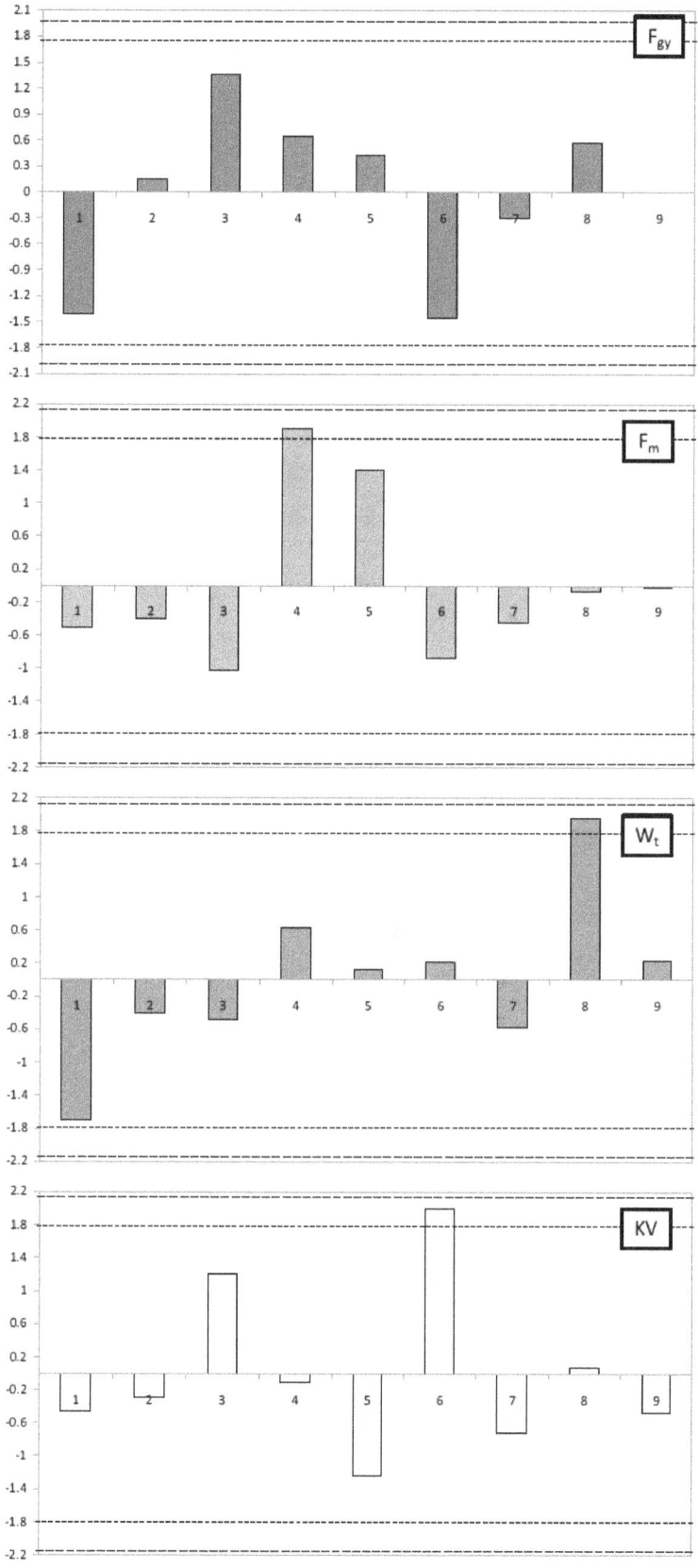

Figure 10 – Mandel's *h* plots for the low-energy specimens.

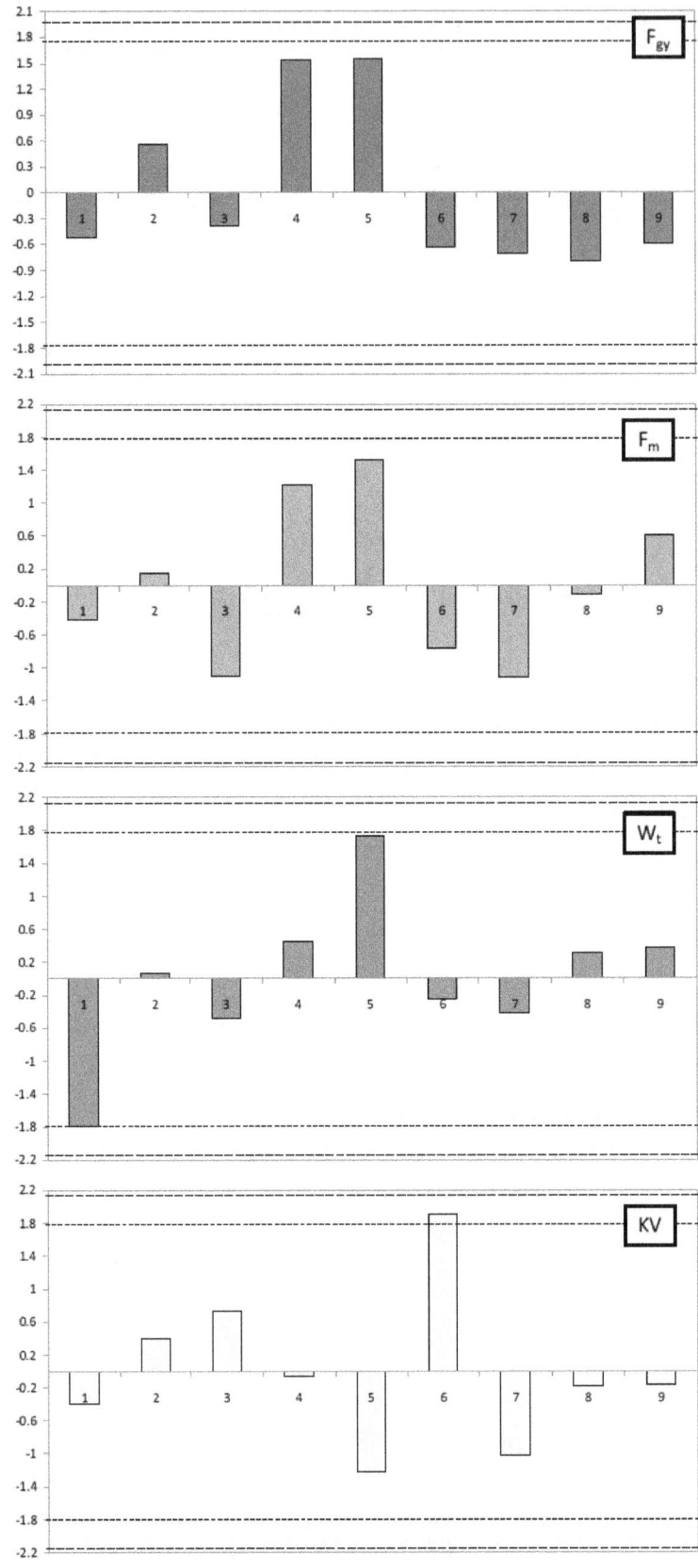

Figure 11 – Mandel's *h* plots for the high-energy specimens.

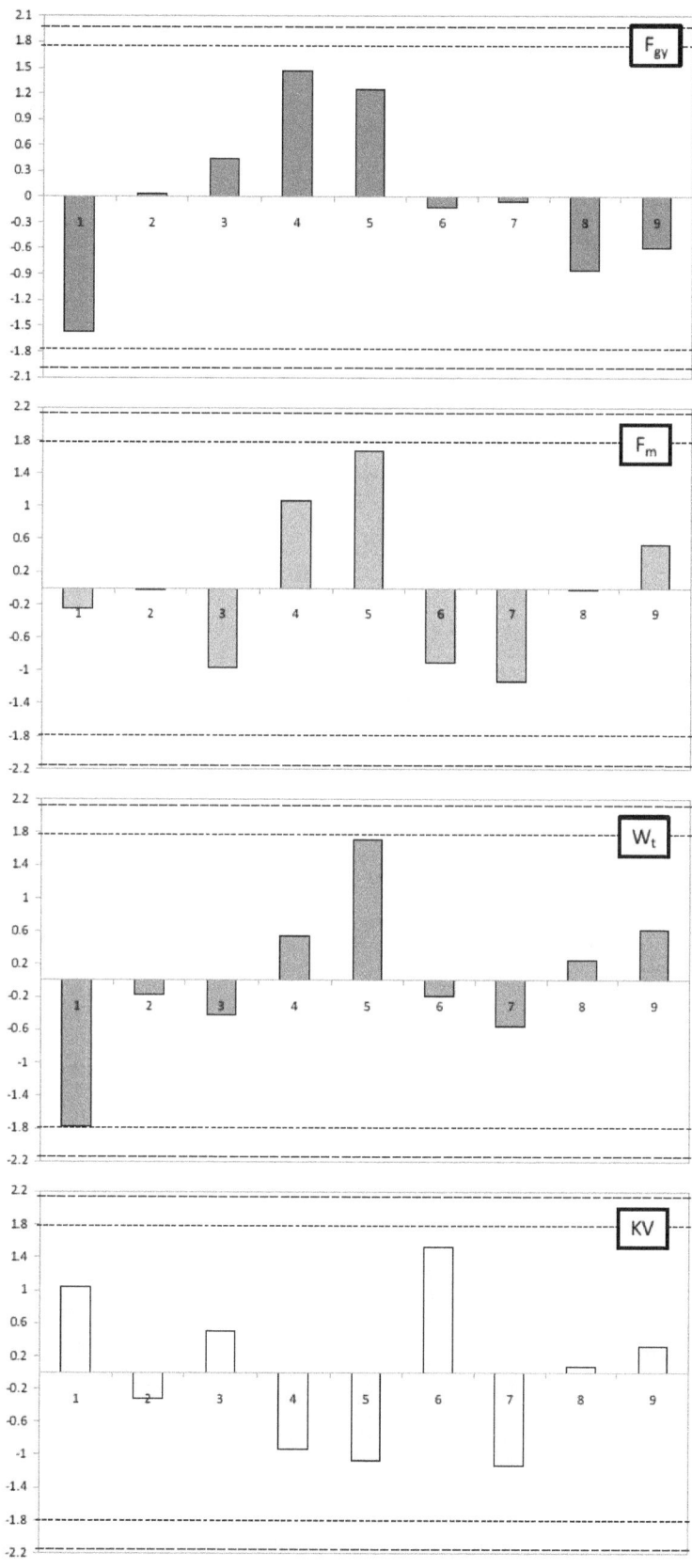

Figure 12 – Mandel's *h* plots for the super-high energy specimens.

17

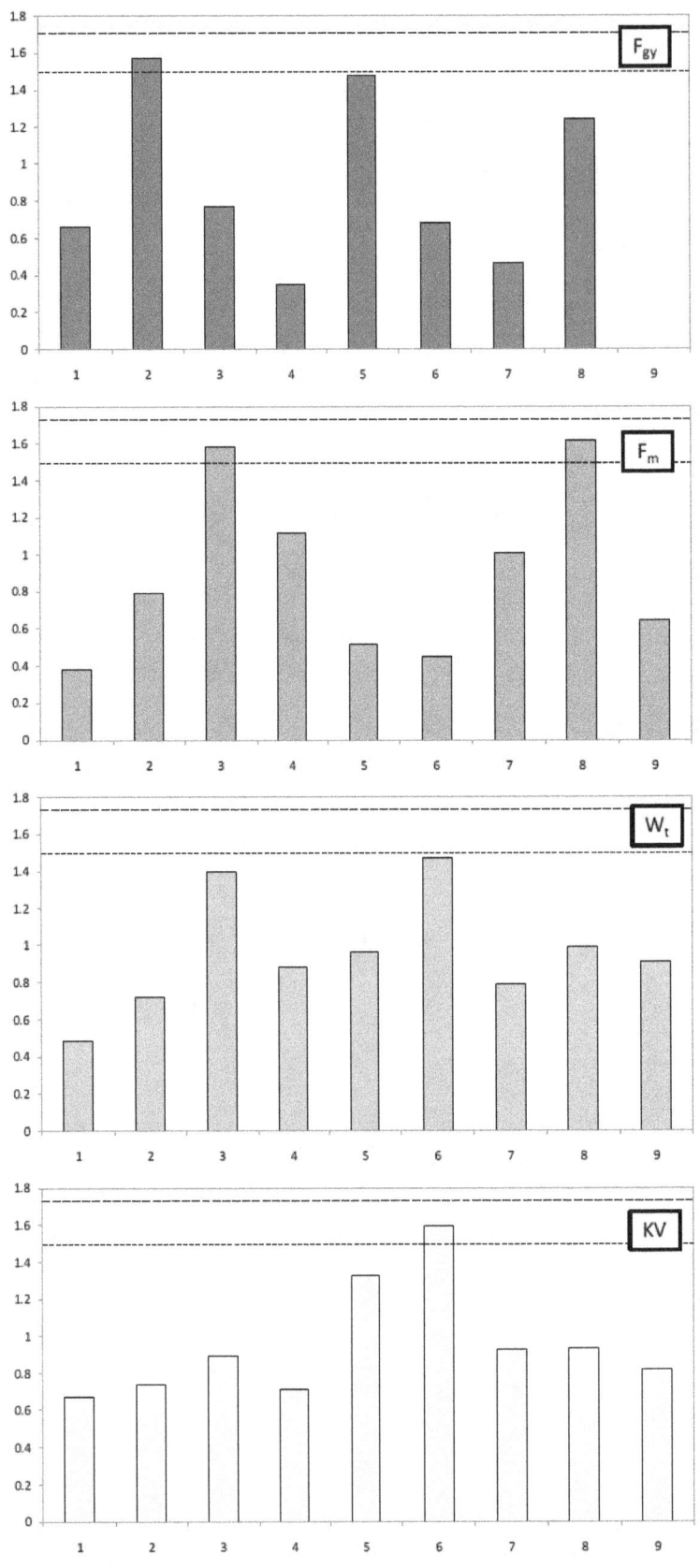

Figure 13 – Mandel's *k* plots for the low-energy specimens.

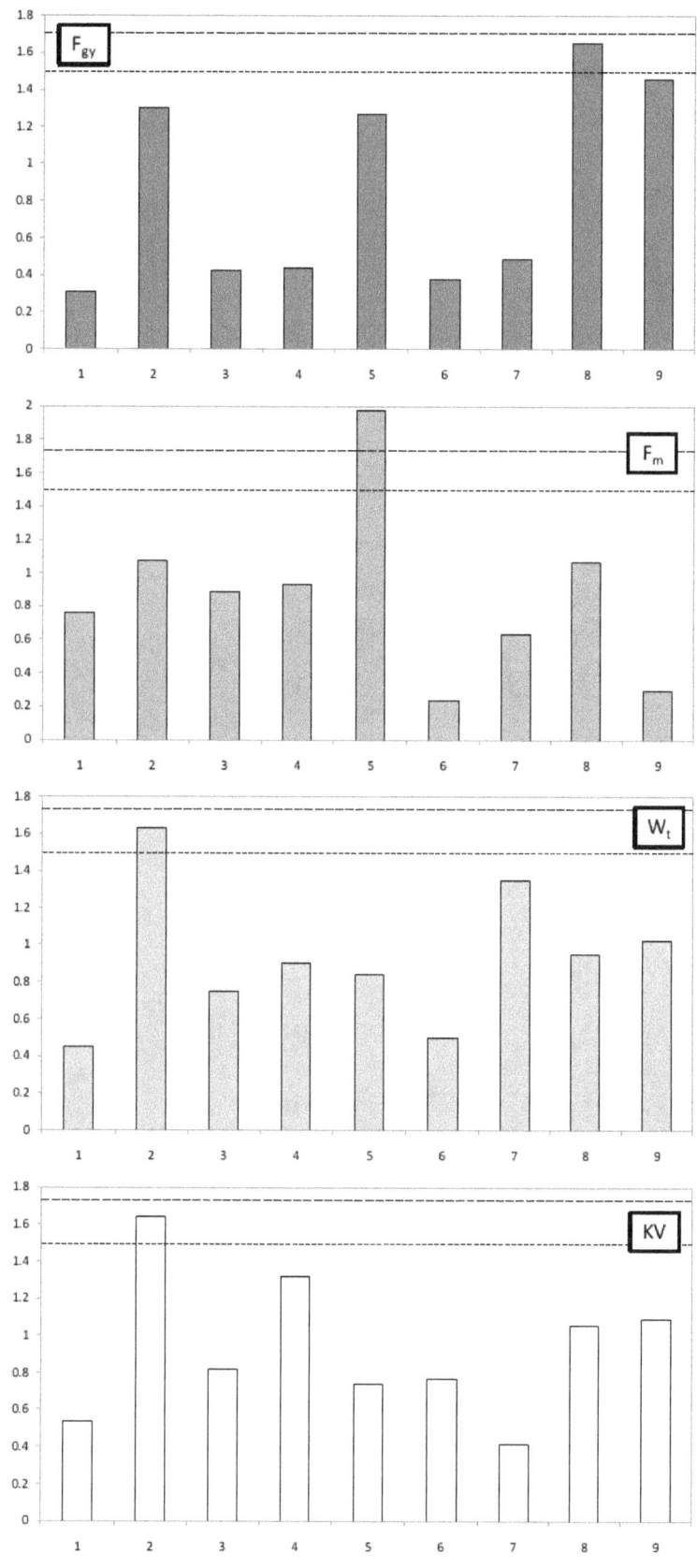

Figure 14 – Mandel's *k* plots for the high-energy specimens.

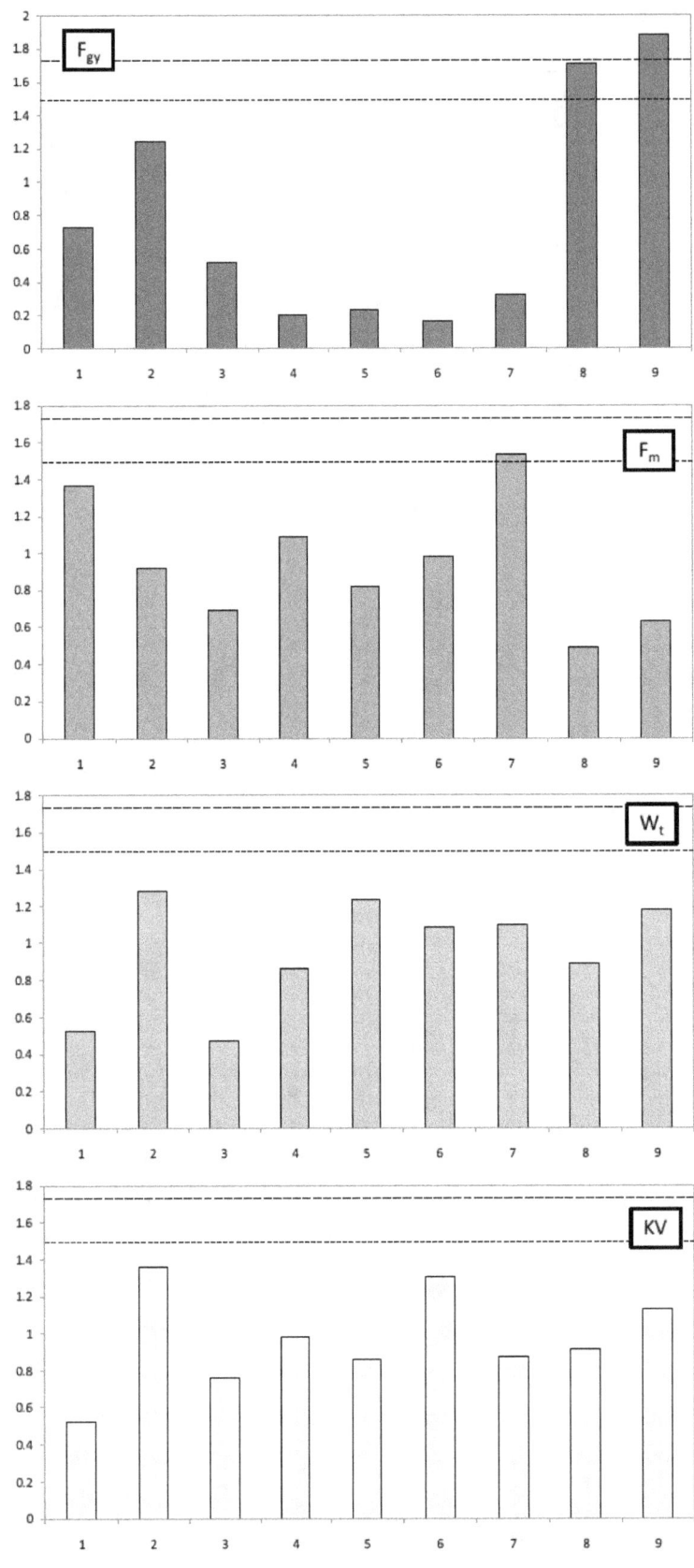

Figure 15 – Mandel's *k* plots for the super-high energy specimens.

20

4.4 Outlier detection

ISO 5725-2:1994 recommends confidence levels of 95 % to define values termed "stragglers" and 99 % to define values termed "statistical outliers".

In the numerical tests used to detect the presence or absence of outliers, it is assumed that the results are distributed in a Gaussian or normal manner, or at least according to a single unimodal distribution. It is also assumed that the number of tests performed and the number of results provided for each parameter are the same for all participants. For the Round-Robin under investigation, these hypotheses are fulfilled with only minor discrepancies (Lab #9 reported no F_{gy} values for the low-energy tests; Lab #5 provided four values for KV instead of five at the low-energy level). This interlaboratory comparison can, therefore, be considered essentially "balanced" and furthermore, results were obtained under repeatability conditions.

For each laboratory or measured parameter, most outlier tests compare some measure of the relative distance of the suspect result from the mean of all results. Many statistical tests are available, but ISO 5725-2:1994 recommends the use of Grubbs' test for lab means and Cochran's test for lab standard deviations.

4.4.1 Grubbs' test

Grubbs' test [5] was used to check whether any of the lab means in Table 7 was exceptionally high or low and would have inflated the estimate of the reproducibility standard deviation, if retained. The statistic used in Grubbs' test is closely related to Mandel's h statistic.

Grubbs' test was performed to establish whether the highest or lowest mean value could be identified as a single straggler or outlier, at the 95 % and 99 % confidence levels, respectively. The results obtained are summarized in Table 9. No stragglers or outliers were identified for any measured parameters at any energy level, i.e. $Z_{max} < Z_{cr,95\%}$.

Table 9 - Results of Grubbs' test on mean values.

Energy level	Parameter	Z_{max}	Lab #	$Z_{cr,95\%}$	$Z_{cr,99\%}$	Outlier/Straggler?
Low	F_{gy} (kN)	1.45	6	2.13	2.28	NO
	F_m (kN)	1.88	4	2.21	2.38	NO
	W_t (J)	1.97	8	2.21	2.38	NO
	KV (J)	1.99	6	2.21	2.38	NO
High	F_{gy} (kN)	1.60	4	2.21	2.38	NO
	F_m (kN)	1.58	5	2.21	2.38	NO
	W_t (J)	1.89	1	2.21	2.38	NO
	KV (J)	2.02	6	2.21	2.38	NO
Super High	F_{gy} (kN)	1.61	1	2.21	2.38	NO
	F_m (kN)	1.77	5	2.21	2.38	NO
	W_t (J)	1.85	1	2.21	2.38	NO
	KV (J)	1.61	6	2.21	2.38	NO

NOTES: Z_{max} = maximum calculated value of Grubbs' statistic; $Z_{cr,95\%}$ and $Z_{cr,99\%}$ are critical values of Grubbs' statistic, corresponding to 95 % and 99 % confidence levels.

4.4.2 Cochran's test

Cochran's test [6] was used to check whether there are standard deviations in Table 8 that were exceptionally large and would have inflated the estimate of the repeatability standard deviation if retained. The statistic used in Cochran's test is closely related to Mandel's k statistic.

This test can identify variances that are greater than the expected variances for the parameter of interest. In this respect, Cochran's test is a one-sided test, as only the laboratory exhibiting the largest variance (but not the one exhibiting the smallest) is tested. Cochran's parameter C is calculated as the ratio between the largest reported variance and the sum of all the variances obtained by participants. The results obtained are summarized in Table 10, where calculated values of Cochran's statistic C are compared to critical values corresponding to 95 % and 99 % confidence limits.

The analyses identified one straggler (Lab #8 – F_{gy} at high-energy level) and three statistical outliers (Lab #5: F_m, high energy; Lab #9: F_{gy}, super-high energy; Lab #7: F_m, super-high energy). The F_m outlier values will be discarded for the calculation of the repeatability standard deviation (section 4.5). It is interesting to note that no stragglers or outliers were found among absorbed energy values (W_t or KV).

Table 10 - Results of Cochran's test on the largest standard deviation for each parameter and energy level.

Energy level	Parameter	C	Lab #	$C_{cr,95\%}$	$C_{cr,99\%}$	Outlier/Straggler?
Low	F_{gy} (kN)	0.31	2	0.391	0.463	NO
	F_m (kN)	0.29	8	0.358	0.425	NO
	W_t (J)	0.24	6	0.358	0.425	NO
	KV (J)	0.28	6	0.358	0.425	NO
High	F_{gy} (kN)	0.40	8	0.358	0.425	STRAGGLER
	F_m (kN)	0.44	5	0.358	0.425	OUTLIER
	W_t (J)	0.30	2	0.358	0.425	NO
	KV (J)	0.30	2	0.358	0.425	NO
Super High	F_{gy} (kN)	0.65	9	0.358	0.425	OUTLIER
	F_m (kN)	12.82	7	0.358	0.425	OUTLIER
	W_t (J)	0.18	2	0.358	0.425	NO
	KV (J)	0.21	2	0.358	0.425	NO

NOTES: C = calculated value of Cochran's statistic; $C_{cr,95\%}$ and $C_{cr,99\%}$ are critical values of Cochran's statistic, corresponding to 95 % and 99 % confidence levels.

4.5 Calculation of repeatability and reproducibility

The precision of the measurement of maximum force F_m and absorbed energy KV was assessed through the analysis of the Round-Robin results in accordance with ASTM E691-12. The procedures in E691 rely on the assumption that an equal number of measurements are taken by each laboratory. Note that Lab #5 reported only 4 values of KV for low-energy specimens (see Table 6). However, this small deviation in sample size is not expected to significantly influence the results of the analysis.

Table 11 lists the grand means[‡‡] for maximum force and absorbed energy for each energy level. Table 12 lists deviations from the grand mean (*i.e.*, the difference between the lab mean and the grand mean) for each participant and material.

[‡‡] The grand mean is the unweighted mean of individual lab means.

Table 11– Grand means for maximum force and absorbed energy.

	Maximum force (kN)	Absorbed energy (J)
Low-energy specimens	2.43	1.59
High-energy specimens	1.79	5.65
Super-high energy specimens	1.79	10.03

Table 12 – Deviations from the grand mean for maximum force and absorbed energy.

Lab #	Low-energy specimens		High-energy specimens		Super-high energy specimens	
	F_m Deviation (kN)	KV Deviation (J)	F_m Deviation (kN)	KV Deviation (J)	F_m Deviation (kN)	KV Deviation (J)
1	-0.10	-0.07	-0.04	-0.06	-0.03	0.21
2	-0.08	-0.05	0.01	0.06	0.00	-0.06
3	-0.20	0.19	-0.11	0.11	-0.10	0.10
4	0.37	-0.02	0.12	-0.01	0.11	-0.19
5	0.28	-0.19	0.15	-0.18	0.18	-0.22
6	-0.17	0.31	-0.08	0.28	-0.09	0.31
7	-0.09	-0.11	-0.11	-0.15	-0.12	-0.23
8	-0.01	0.01	-0.01	-0.03	0.00	0.02
9	0.00	-0.07	0.06	-0.03	0.06	0.06

The calculations of Mandel's h and k statistics are detailed in section 4.3, and illustrated in Figures 10 to 15. The individual values of h and k calculated for each lab are collected in Table 13 (F_m) and Table 14 (KV).

Table 13 – Laboratory consistency statistics for maximum forces. The red bold value exceeds the 99 % critical value.

Lab #	Low-energy specimens		High-energy specimens		Super-high energy specimens	
	h	k	h	k	h	k
1	-0.49597	0.37931	-0.43486	0.76357	-0.25958	1.36639
2	-0.39675	0.79050	0.14819	1.07248	-0.02034	0.92122
3	-1.01766	1.57956	-1.13956	0.88791	-1.00523	0.69224
4	1.90279	1.11271	1.27025	0.93303	1.11807	1.08532
5	1.40667	0.51359	1.57855	1.97921	1.75407	0.82396
6	-0.87445	0.44702	-0.79141	0.23379	-0.93745	0.98071
7	-0.43766	1.00905	-1.15948	0.63430	-1.18507	1.53345
8	-0.07177	1.61266	-0.10645	1.06885	-0.02232	0.48891
9	-0.01520	0.64470	0.63476	0.30183	0.55784	0.62931

Table 14 – Laboratory consistency statistics for absorbed energies.

Lab #	Low-energy specimens		High-energy specimens		Super-high energy specimens	
	h	k	h	k	h	k
1	-0.44848	0.67190	-0.41867	0.53384	1.09192	0.52008
2	-0.29336	0.73706	0.42623	1.64387	-0.33438	1.35728
3	1.20622	0.89206	0.77585	0.81975	0.52973	0.76121
4	-0.11237	0.71064	-0.05449	1.31887	-0.99027	0.98421
5	-1.23706	1.32339	-1.29271	0.73631	-1.12561	0.86294
6	1.99480	1.59704	2.02863	0.76825	1.61247	1.31095
7	-0.71065	0.92555	-1.08629	0.41705	-1.20401	0.87835
8	0.07524	0.93688	-0.19296	1.05254	0.08822	0.91293
9	-0.47434	0.81746	-0.18559	1.08957	0.33192	1.12738

None of the participating labs provided h values higher than 2.23, which is the critical value at the 0.5 % significance level for $p = 9$ (number of labs). This is in agreement with the outcome of the Grubbs' test documented in section 4.4.1. Therefore, none of the lab means should be removed from the analyses on the grounds of inconsistency with respect to the other labs.

One of the labs (Lab #5, F_m at the high-energy level) returned a k value higher than 1.81, which is the critical value at the 0.5 % significance level for $p = 9$ and $n = 5$ (the number of measurements per lab). Similarly, Cochran's test documented in section 4.4.2 classifies the same data set as a statistical outlier. Therefore, this data set and its standard deviation will be removed from the calculation of the repeatability standard deviation S_r.

The precision statistics calculated for each parameter and each energy level are given in Table 15: grand mean[§§], repeatability standard deviation (S_r), 95 % repeatability limit ($r = 2.8S_r$), reproducibility standard deviation (S_R), and 95 % reproducibility limit ($R = 2.8S_R$).

Table 15 – Precision statistics calculated from the Round-Robin results.

Energy level	Parameter	S_r	r	S_R	R
Low	F_m (kN)	0.035	0.099	0.196	0.547
	KV (J)	0.063	0.175	0.155	0.433
High	F_m (kN)	0.019	0.052	0.095	0.267
	KV (J)	0.109	0.305	0.137	0.384
Super High	F_m (kN)	0.013	0.037	0.100	0.281
	KV (J)	0.226	0.634	0.226	0.634

4.6   Additional analyses performed on force and absorbed energy values

4.6.1   Relationship between different measures of absorbed energies (W_t and KV)

It has been contended [7,8] that the difference between the absorbed energy returned by the machine dial/encoder (KV) and that calculated from the instrumented force-displacement curve (W_t) is a good indicator of the overall performance of an instrumented impact tester. Specifically, ASTM E2298-13 requires the difference between KV and W_t to be within 15 % of KV or 1 J, whichever is larger. This corresponds to the following limits for the tests under investigation: 1 J at the low- and high-energy level, and 1.5 J (15 % of the grand mean of KV) at the super-high energy level.

The relationship between the two measures of absorbed energy varies from machine to machine, and depends on the characteristics of the instrumented striker (design and static calibration) [7,8].

For the nine laboratories participating in the Round-Robin, Table 16 shows for each test performed the relationship between KV and W_t in terms of the difference $|KV - W_t|$ for low and high energies, and the percent difference relative to W_t for super-high energies.

[§§] The grand mean is not the same as the certified reference value calculated in section 5.

24

Table 16 – Relationship between different measures of absorbed energies (average values).

Lab #	Low energy $\|KV\text{-}W_t\|$ (J)	High energy $\|KV\text{-}W_t\|$ (J)	Super-high energy $100 \times \left\|\dfrac{KV}{W_t} - 1\right\|$
	0.23	0.68	14.9%
	0.21	0.72	14.0%
1	0.20	0.72	13.8%
	0.19	0.72	15.0%
	0.18	0.70	13.6%
	0.12	0.04	0.9%
	0.12	0.01	0.3%
2	0.10	0.11	0.0%
	0.12	0.04	0.3%
	0.11	0.10	0.4%
	0.29	0.42	1.5%
	0.35	0.25	5.0%
3	0.39	0.34	4.4%
	0.39	0.31	4.6%
	0.34	0.35	2.1%
	0.05	0.14	4.3%
	0.06	0.14	5.5%
4	0.08	0.21	5.3%
	0.04	0.14	5.4%
	0.03	0.21	5.1%
	0.09	0.91	13.6%
	0.06	0.89	10.7%
5	0.05	0.85	10.8%
	0.09	0.80	11.5%
	-	0.88	10.8%
	0.35	0.41	4.3%
	0.57	0.37	3.5%
6	0.36	0.43	4.4%
	0.42	0.42	4.5%
	0.37	0.42	4.1%
	0.08	0.21	2.1%
	0.06	0.09	1.0%
7	0.05	0.14	0.8%
	0.05	0.11	1.8%
	0.07	0.13	0.6%
	0.03	0.14	1.6%
	0.04	0.13	1.3%
8	0.03	0.12	1.8%
	0.03	0.12	1.3%
	0.03	0.13	1.3%
	0.03	0.18	3.1%
	0.05	0.10	3.4%
9	0.03	0.16	2.6%
	0.01	0.12	3.1%
	0.03	0.19	3.3%

All the absolute deviations are smaller than 1 J for low and high energy, while all the percent differences are below 15 % for super-high energy; therefore, all the machines participating in the Round-Robin satisfy the ASTM E2298-13 requirements.

Values of W_t and KV, with bounds corresponding to the ASTM E2298-13 limits, are illustrated in Figure 16 (low energy), Figure 17 (high energy), and Figure 18 (super-high energy). The limits are 1 J at the low- and high-energy levels, and approximately 1.5 J (15 % of the mean KV) at the super-high energy level. (ASTM E2298-13 specifies that differences are evaluated for each pair of measurements, so the limits based on the mean KV in Figure 18 are approximate.) In Figure 18, there is one value for Lab #5 lying outside the approximate

ASTM limits. However, the percent difference calculated for this point (13.6 %), based on individual measurements of KV and W_t, is smaller than 15 % of W_t.

Examination of Figures 16 through 18 shows that trends from individual laboratories for KV and W_t are consistent across the different energy levels, thus confirming that the relationship between the two parameters is machine- and striker-dependent. Based on these comparisons, the overall performance of the participants' equipment appears satisfactory.

It is also worth noting that at the high-energy level (Figure 17) and at the super-high energy level (Figure 18), instrumented energy values show much more scatter (horizontal spread) than encoder energy values, as should be expected due to the additional uncertainty introduced by individual striker calibrations. This effect is less evident for low-energy data (Figure 16).

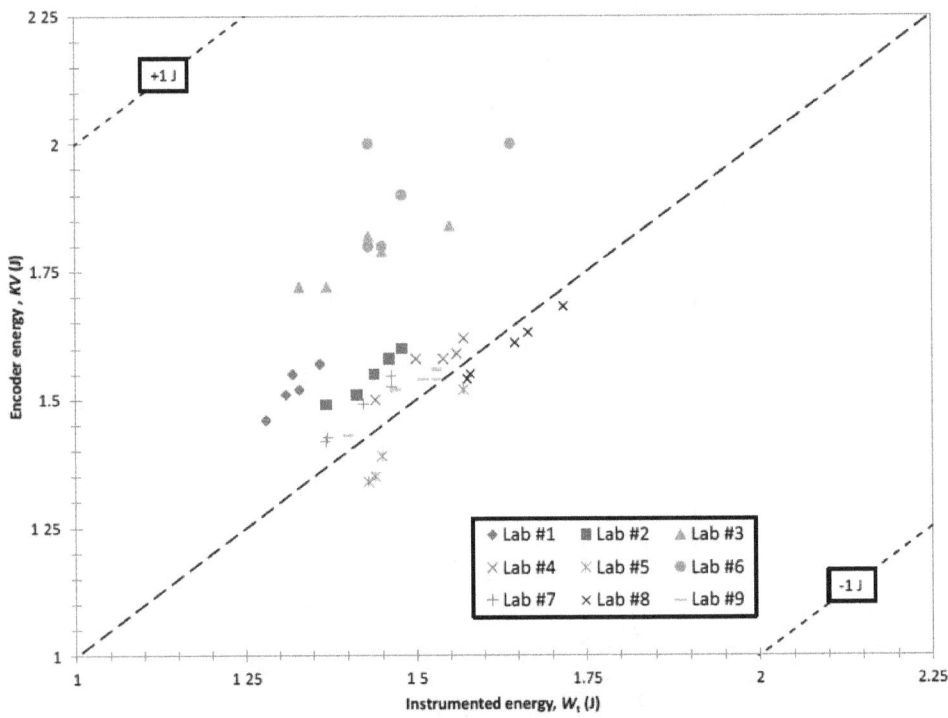

Figure 16 - Instrumented and encoder absorbed energy values from low-energy KLST verification specimens.

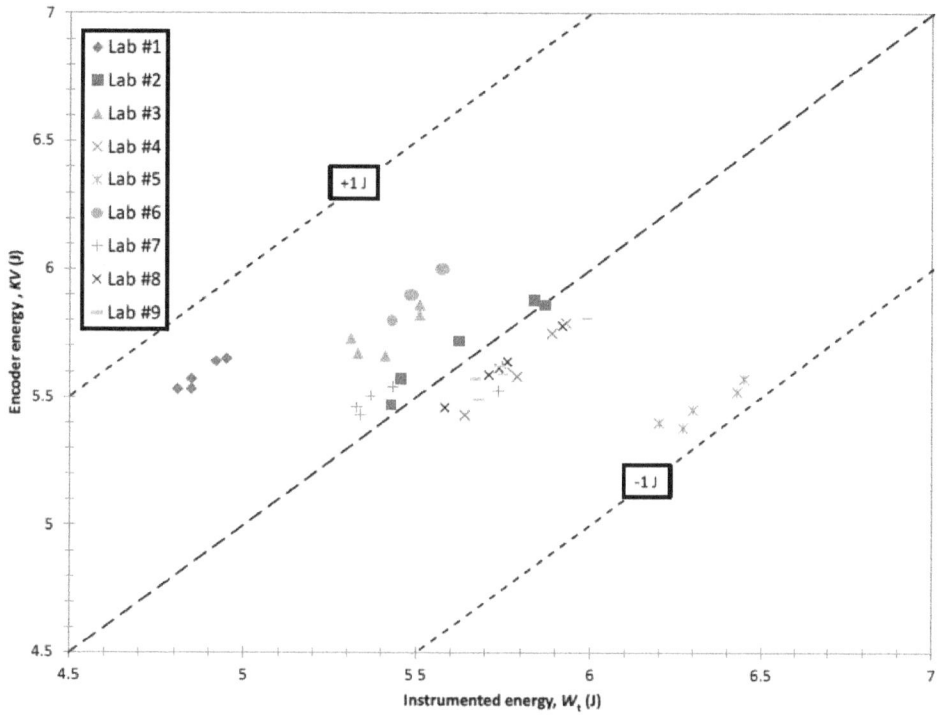

Figure 17 - Instrumented and encoder absorbed energy values from high-energy KLST verification specimens.

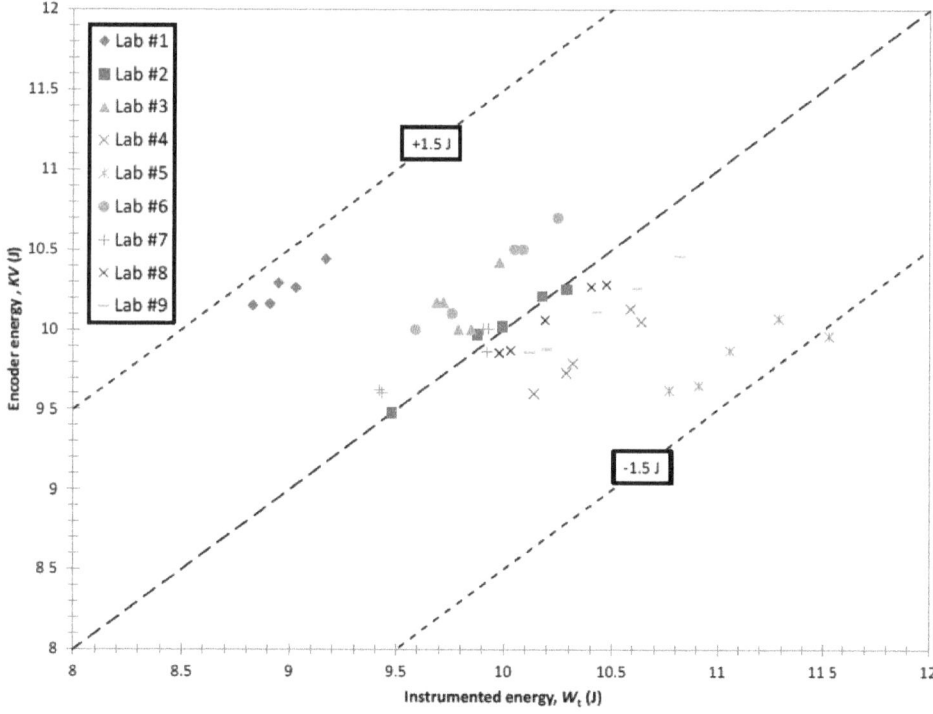

Figure 18 - Instrumented and encoder absorbed energy values from super-high energy KLST verification specimens.

4.6.2 Correlation between full-size and KLST absorbed energies

The correlations between values of KV yielded by different Charpy specimen types have been extensively investigated. An overview of various correlations of upper shelf energy (USE) data was provided by Sokolov and Alexander [9]. An additional correlation approach is available in [10].

The method commonly used in Europe consists of establishing an empirical ratio between full-size USE (USE_{fs}) and sub-size USE (USE_{ss}) based on a large number of tests. A different approach, often used by US and Japanese researchers, correlates USE values with the ratio of various geometrical parameters, GP_x (with $x = fs$ or ss) for different specimen geometries in the form:

$$\frac{USE_{fs}}{USE_{ss}} = \frac{GP_{fs}}{GP_{ss}} \quad . \tag{1}$$

Eq. (1) can also be expressed in terms of a normalization factor NF, which corresponds to the ratio of geometrical parameters mentioned above:

$$USE_{fs} = NF \times USE_{ss} \quad . \tag{2}$$

The most common expressions for NF that can be found in the literature are the following:

$$NF_1 = \frac{(Bb)_{fs}}{(Bb)_{ss}} \quad , \tag{3}$$

based on the ratio of fracture areas, with B = specimen thickness and b = ligament size [11,12];

$$NF_2 = \frac{\left[(Bb)^{3/2}\right]_{fs}}{\left[(Bb)^{3/2}\right]_{ss}} \quad , \tag{4}$$

based on the ratio of nominal fracture volumes [11,12];

$$NF_3 = \frac{(Bb^2)_{fs}}{(Bb^2)_{ss}} \quad , \tag{5}$$

based on a different expression for the ratio of nominal fracture volumes [13,14], and

$$NF_4 = \frac{\left(\dfrac{Bb^2}{LK_t}\right)_{fs}}{\left(\dfrac{Bb^2}{LK_t}\right)_{ss}} \quad , \tag{6}$$

where L = span and K_t = elastic stress concentration factor [15].

In addition, Sokolov and Alexander [9] established empirical normalization factors NF_5 for 4 types of sub-size specimens considered in their study (one of which corresponds to KLST) by averaging the values of USE_{fs}/USE_{ss} obtained on ten different materials (mostly pressure vessel steels with different heat treatments).

In [10], an exponential relationship between full-size and KLST values of *USE* was established, as shown in Figure 19. The best-fit regression curve relating USE_{fs} and USE_{KLST} is given by

$$USE_{fs} = 29.454\, e^{0.2378 \cdot USE_{KLST}} \qquad . \qquad (7)$$

Note that eq.(7) is strictly applicable only in case of upper-shelf behavior, and within the USE_{KLST} range shown in Figure 19.

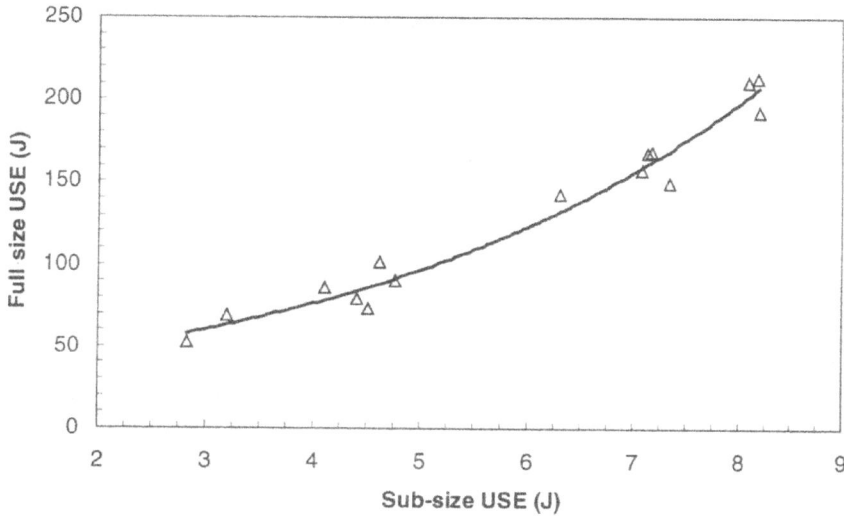

Figure 19 - Exponential correlation between sub-size and full-size *USE* values for KLST specimens [10].

Finally, the average value of USE_{fs}/USE_{KLST} calculated in [10] for nine unirradiated and irradiated pressure vessel steels was $NF_7 = 21.6$. A summary of the different normalization factors available for KLST miniaturized specimens, according to the methods listed above, is provided in Table 17.

Table 17 - Summary of normalization factors for estimating full-size *USE* based on KLST *USE*, see eq. (3). Note: the value of NF_6, which would result from Eq.(7) [10], is not shown in the Table because it depends on USE_{fs}.

NF_1 [Eq.(3)]	NF_2 [Eq.(4)]	NF_3 [Eq.(5)]	NF_4 [Eq.(6)]	NF_5 [9]	NF_7 [10]
8.9	26.5	23.7	13	24.9	21.6

Considering the KLST tests performed at three energy levels in the framework of the Round-Robin, experimental normalization factors NF_{exp} were calculated by dividing the certified/average values of KV_{fs} at room temperature by the average KLST absorbed energies for every data set. The results are shown in Table 18 and should be compared with the normalized factors listed in Table 17.

Table 18 - Experimental normalization factors obtained from KLST tests performed by Round-Robin participants.

Energy level	Specimen type	$\overline{KV}$ (J)	$NF_{exp} = \dfrac{\overline{KV}_{fs}}{\overline{KV}_{ss}}$
Low	Full-size KLST	18.2 (cv) 1.60 (mv)	11.4
High	Full-size KLST	105.3 (cv) 5.64 (mv)	18.7
Super-high	Full-size KLST	239.8 (mv) 10.05 (mv)	23.8

NOTE – cv: certified value (at RT); mv = mean value (at RT).

Theoretical (Table 17), empirical (Figure 19) and experimental (Table 18) normalization factors are compared in Figure 20.

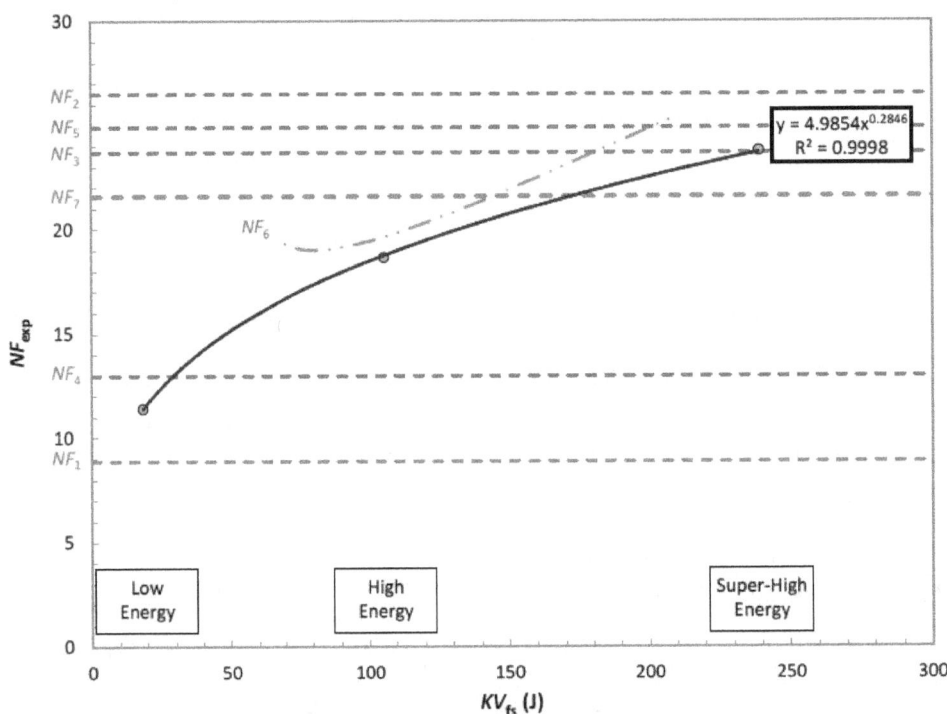

Figure 20 - Comparison between theoretical, empirical and experimental normalization factors for KLST specimens at different energy levels.

Another approach for evaluating the different normalization factors is shown in Figure 21, which compares certified/average values of KV_{fs} to predicted values obtained from KV_{KLST} by the use of the seven normalization factors presented above. Note that in Figure 21, NF_6 from Eq.(7) has been applied only to the high-energy level (KV_{KLST} within the applicability range).

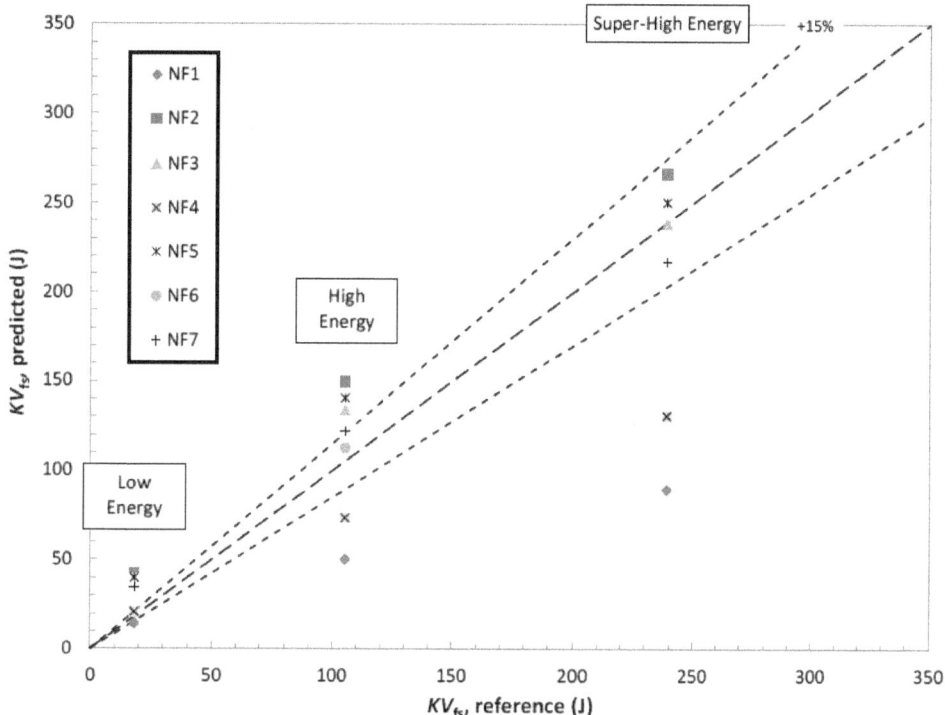

Figure 21 - Prediction of full-size absorbed energy from KLST specimens.

When examining Figures 20 and 21, one must remember that all the methods described above address *USE* values, rather than generic values of absorbed energy. At the high- and super-high energy levels, analysis of the instrumented traces indicates fully ductile behavior; therefore it is appropriate to assume $KV = USE$. For low-energy specimens, however, the material's behavior is typical of the ductile-to-brittle transition regime (shear fracture appearance values, both measured by some participants and estimated from the analysis of the instrumented test records, range between 27 % and 71 %), and therefore it is not appropriate to assume $KV = USE$.

The most immediate conclusion is that normalization factors are dependent of absorbed energy. In particular, NF_1 and NF_4 work acceptably only at the low-energy level, while at higher energies they significantly underpredict KV_{fs}. Conversely, non-conservative estimates were generally obtained through the use of NF_2. Mixed results are observed for NF_3, NF_5, NF_6, and NF_7. Equation (7) works only at the high-energy level, *i.e.*, when both KV_{fs} and KV_{KLST} are within the ranges for which the relationship was developed (Figure 19). In short, none of the approaches considered appears convincing for all energy levels.

Based on test results, the following relationship between KV_{fs} and KV_{KLST} was obtained (Figure 22):

$$KV_{fs} = 9.448 \cdot KV_{KLST}^{1.3979} \qquad . \tag{8}$$

Eq. (8) will be used to normalize the value 1.4 J for the calculation of sample sizes in section 6.

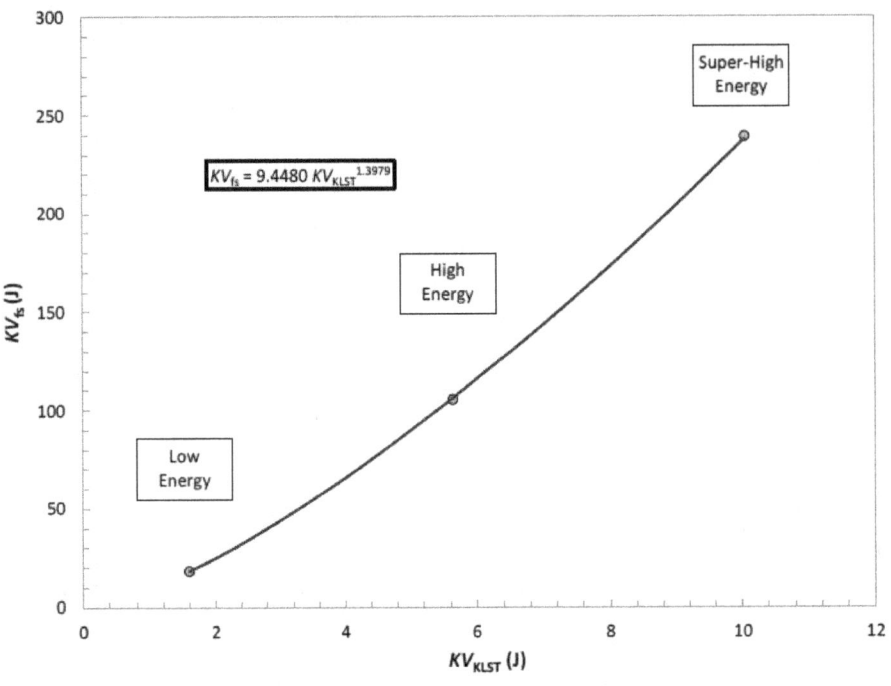

Figure 22 – Correlation obtained between KV_{KLST} and KV_{fs}.

4.6.3 Correlation between full-size and KLST maximum forces

The relationship observed between the mean values of maximum forces measured from KLST specimens and the corresponding reference/average values for full-size specimens is shown in Figure 23. General trends were found to be consistent between the two specimen types, *i.e.* the highest F_m was observed at low energy; super-high energy specimens provide slightly higher F_m values than high-energy specimens.

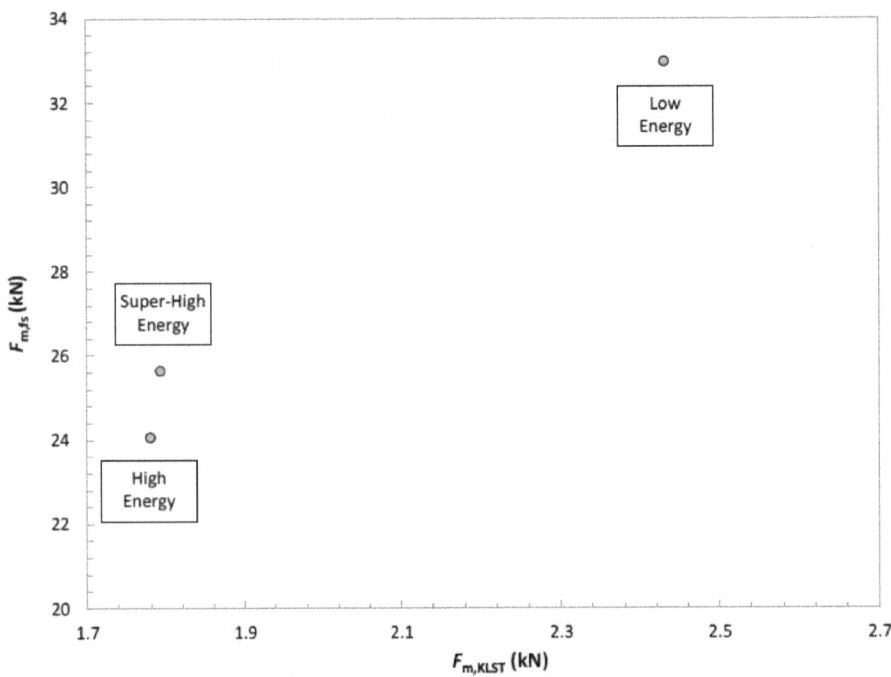

Figure 23 – Correlation obtained between $F_{m,KLST}$ and $F_{m,fs}$.

4.7 Optional results reported

4.7.1 Lateral expansion

Four of the Round-Robin participants (Labs #3, 5, 6, 9) reported values of lateral expansion. The mean values are presented in Table 19.

Table 19 – Average values of lateral expansion reported by four participants.

Lab #	Low-energy Lat. Exp. (mm)	High-energy Lat. Exp. (mm)	Super-high energy Lat. Exp. (mm)
3	0.07	0.56	1.00
5	0.06	0.49	0.92
6	0.06	0.45	0.74
9	0.10	0.56	1.04

The relationship between reported values of absorbed energy and lateral expansion is shown in Figure 24.

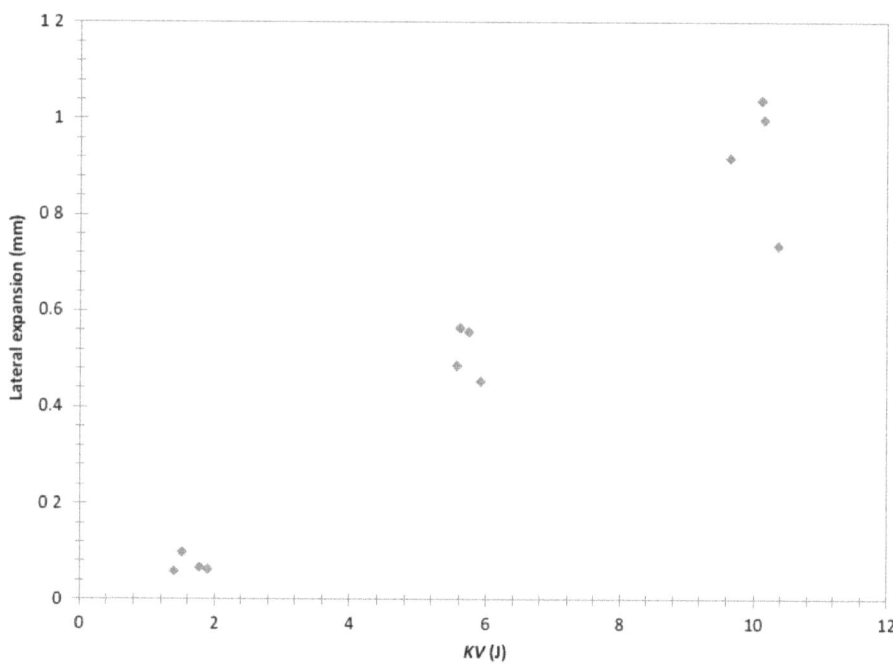

Figure 24 – Relationship between absorbed energy and lateral expansion for KLST specimens.

4.7.2 Shear Fracture Appearance

Only three Round-Robin participants (Labs #1, 8, 9) measured shear fracture appearance (*SFA*) on the low-energy KLST specimens tested. In addition, *SFA* values were estimated[***] from characteristic force values with the following relationship [16]:

$$SFA_{est} = 1 - \frac{F_{iu} - F_a}{F_m + 0.5(F_m - F_{gy})} \tag{9}$$

[***] These calculations were performed by the Round-Robin coordinator.

33

Average *SFA* values (measured and estimated) for Round-Robin participants are shown in Table 20 and illustrated in Figure 25.

Table 20 – Average values of SFA (measured and estimated).

Lab #	SFA_{meas} (%)	SFA_{est} (%)
1	42	58
2	-	64
3	-	27
4	-	37
5	-	68
6	46	49
7	-	71
8	-	67
9	48	62

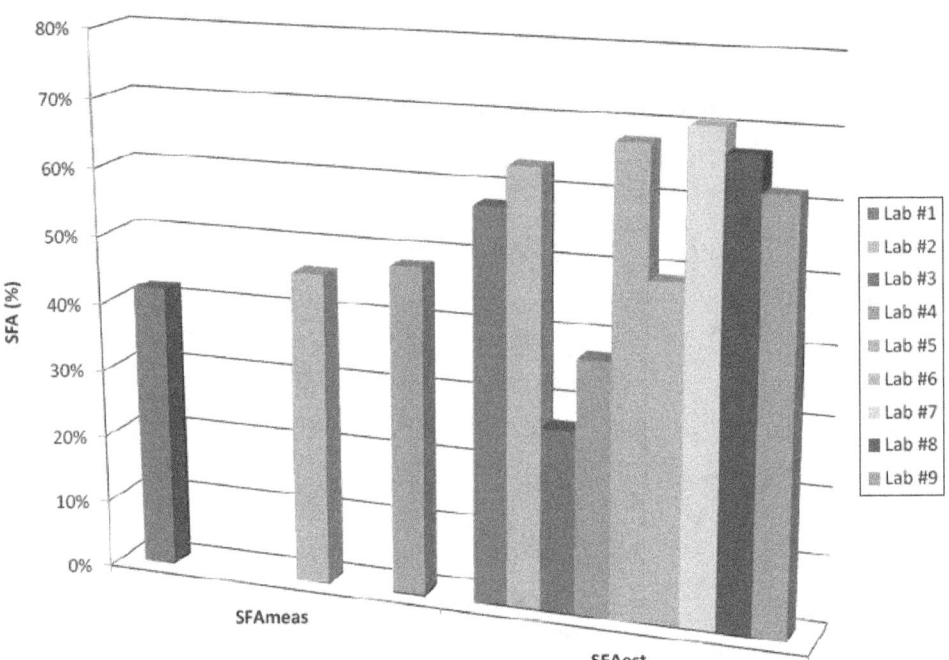

Figure 25 – Measured and estimated *SFA* values for low-energy specimens.

Figure 26 shows that Eq. (9) tends to overestimate measured values of *SFA*.

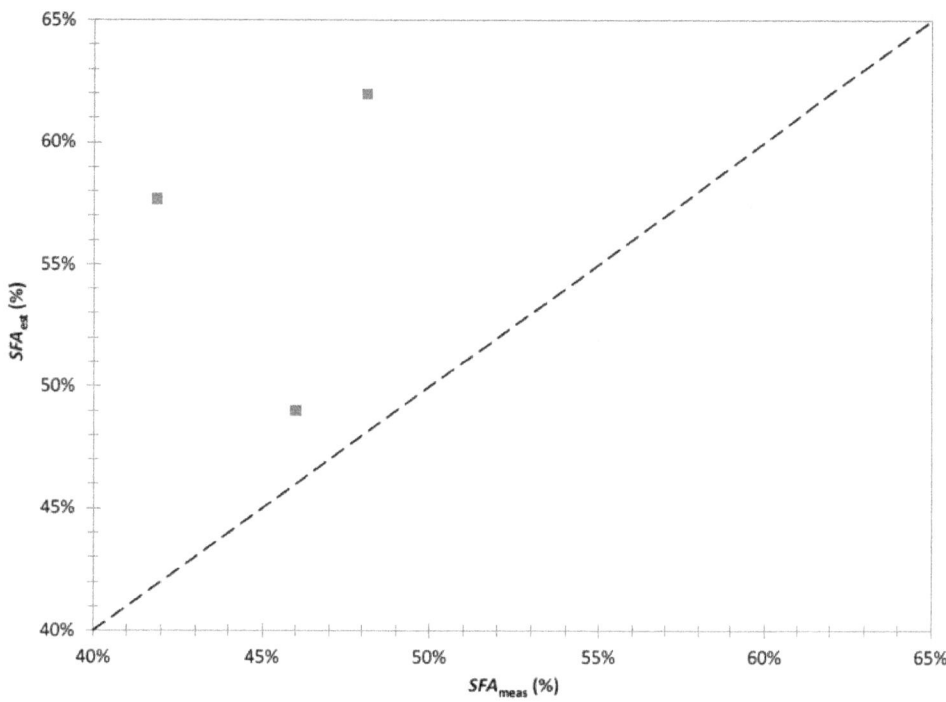

Figure 26 – Relationship between measured and estimated *SFA* values for low-energy specimens.

5.   Establishment of certified reference values

The certified reference values for maximum force F_m and absorbed energy KV were determined from the Round-Robin results by the use of the Mandel-Paule method for computing consensus values [17].

Consensus values depend on the assumption that all impact testers in the study are "good" and no single machine is expected to provide better results than the others. A consensus value is a weighted mean in which the weights account for both within-machine and between-machine variation.

The certified reference value (consensus value) $\tilde{Y}$ is computed from:

$$\tilde{Y} = \frac{\sum_{i=1}^{k} \omega_i \bar{Y}_i}{\omega_i},\tag{10}$$

where k is the number of machines in the study, ω_i is the weight for the ith machine, and $\bar{Y}_i$ is the average of the measurements for the ith machine. The standard uncertainty of the certified reference value is:

$$u(\tilde{Y}) = \frac{1}{\sqrt{\sum_{i=1}^{k} \omega_i}},\tag{11}$$

and the estimated weight for the ith machine is:

$$\omega_i = \frac{1}{\frac{s_i^2}{n_i} + s_b^2},\tag{12}$$

where s_i^2 is the sample variance of the ith machine (s_i values are given in Table 8), n_i is the number of test results for the ith machine, and s_b^2 is the estimated variance between machines, which is established by an iterative procedure.

The calculations performed are summarized in Table 21, which also provides additional calculated parameters (unweighted mean of the means). Further details of the calculations performed are provided in Annex B.

Table 21 - Results of the calculations performed for the establishment of consensus values.

Energy level	Parameter	Consensus value, $\tilde{Y}$	Between-lab variance, s_b^2	Mean of the means, $\bar{Y}$	Standard uncertainty, $u(\tilde{Y})$
Low	F_m (kN)	2.43	0.038	2.43	0.196
	KV (J)	1.59	0.022	1.59	0.152
High	F_m (kN)	1.78	0.011	1.79	0.105
	KV (J)	5.65	0.014	5.65	0.121
Super High	F_m (kN)	1.79	0.010	1.79	0.101
	KV (J)	10.03	0.041	10.03	0.221

For the calculation of expanded uncertainties corresponding to 95 % uncertainty intervals, a coverage factor is used. This is based on a t-table value for $9 - 1 = 8$ degrees of freedom and a 95 % two-sided uncertainty interval. The values obtained are summarized in Table 22.

Table 22 – Certified reference values, standard uncertainties, coverage factors, and expanded uncertainties.

Energy level	Parameter	Certified reference value	Standard uncertainty	Coverage factor, k	Expanded uncertainty
Low	F_m (kN)	2.43	0.065	2.306	0.150
	KV (J)	1.59	0.051	2.306	0.117
High	F_m (kN)	1.78	0.035	2.306	0.081
	KV (J)	5.65	0.043	2.306	0.098
Super High	F_m (kN)	1.79	0.034	2.306	0.077
	KV (J)	10.03	0.075	2.306	0.173

6.   Assessment of sample size

According to the standard operating procedures of the Charpy Verification Program at NIST [2], the sample size n is the minimum number of specimens from a given production lot that should be tested in a verification test. It is calculated as:

$$n = \left(\frac{3s_p}{E}\right)^2, \tag{13}$$

where s_p is the pooled standard deviation (or the machine standard deviation if only one machine is used), and E is the greater of 0.255 J or 5 % of the certified reference value. The derivation of $E = 0.255$ J is obtained by normalizing 1.4 J as described hereunder.

The NIST procedure for standard, full-size Charpy specimens [2] uses the greater of 1.4 J and 5 % of KV for the factor E in Eq. (13). For KLST specimens, the value 1.4 J was converted by means of the exponential regression function of Eq. (8).

The results of the sample-size calculations based on KV data reported in the Round-Robin are given in Table 23.

Table 23 – Sample-size calculations for KLST verification specimens.

Energy level	s_p (J)	5 % $\overline{KV}$ (J)	E (J)	Sample size, n
Low	0.062	0.080	0.255	0.5
High	0.109	0.282	0.282	1.3
Super-High	0.226	0.503	0.503	1.8

The values of n reported in Table 23 would justify, from a purely statistical standpoint, a sample size of 2 specimens in a set for SRM 2216, SRM 2218, and SRM 2219. A sample size larger than 5 would indicate that the material in question is not adequate for producing reference specimens.

Acknowledgements

The collaboration of all the laboratories participating in the Round-Robin is gratefully acknowledged.

Bibliography

[1] C. McCowan, R. Santoyo, and J. Splett, "Certification Report for SRMs 2112 and 2123," NIST Special Publication 260-172, July 2009.

[2] C. N. McCowan, T. A. Siewert, and D. P. Vigliotti, "The NIST Charpy V-notch Verification Program: Overview and Operating Procedures," in: NIST Technical Note 1500-9, *Charpy Verification Program: Reports Covering 1989-2002*, September 2003, pp. 3-42.

[3] C. N. McCowan, E. Lucon, and R. L. Santoyo, "Evaluation of Bias for Two Charpy Impact Machines with the Same Instrumented Striker," *Journal of ASTM International*, Vol. 8, No. 5, 2011.

[4] J. Mandel, "A new analysis of interlaboratory test results," in *ASQC Quality Congress Transaction –* Baltimore, 1985, pp. 360-366.

[5] F. E. Grubbs, "Sample Criteria for Testing Outlying Observations," *Annals of Mathematical Statistics*, Vol. 21, Number 1, 1950, pp. 27-58.

[6] W.G. Cochran, "The distribution of the largest of a set of estimated variances as a fraction of their total," *Annals of Human Genetics*, London, Vol. 11, Number 1, Jan 1941, pp. 47–52.

[7] E. Lucon, R. Chaouadi, and E. van Walle, "Different Approaches for the Verification of Force Values Measured with Instrumented Charpy Strikers," in *Pendulum Impact Machines: Procedures and Specimens*, ASTM STP 1476 (Eds: T. Siewert, M. Manahan, and C. McCowan), ASTM, West Conshohocken, PA, 2006, pp. 95-102.

[8] E. Lucon, "On the Effectiveness of the Dynamic Force Adjustment for Reducing the Scatter of Instrumented Charpy Results," *Journal of ASTM International*, Vol. 6, No. 1 (January 2009).

[9] M. A. Sokolov and D. J. Alexander, "An Improved Correlation Procedure for Subsize and Full-Size Charpy Impact Specimen Data," NUREG/CR-6379, ORNL-6888, March 1997.

[10] E. Lucon, R. Chaouadi, A. Fabry, J.-L. Puzzolante, and E. van Walle, "Characterisation of materials properties by use of full size and subsize Charpy tests: an overview of different correlation procedures," in Pendulum Impact Testing: A Century of Progress, ASTM STP 1380 (Ed: T.A. Siewert), ASTM, Philadelphia, PA 2000, pp. 146-163.

[11] W. R. Corwin, R. L. Klueh, and J. M. Vitek, "Effect of specimen size and nickel content on the impact properties of 12 Cr-1 MoVW ferritic steel," *Journal of Nuclear Materials* Vol. 122, Issues 1-3, 1984, pp. 343-348.

[12] W. R. Corwin, and A. M. Hougland, "Effect of Specimen Size and Material Condition on the Charpy Impact Properties of 9Cr-1Mo-V-Nb Steel," in *The Use of Small-Scale Specimens for Testing Irradiated Material*, ASTM STP 888 (Eds: W. R. Corwin, G. E. Lucas), ASTM, Philadelphia, PA 1986, pp. 325-338.

[13] G. E. Lucas, G. R. Odette, J. W. Sheckherd, P. McConnell, and J. Perrin, "Subsized Bend and Charpy V-Notch Specimens for Irradiated Testing," in *The Use of Small-Scale Specimens for Testing Irradiated Material*, ASTM STP 888 (Eds: W. R. Corwin, G. E. Lucas), ASTM, Philadelphia, PA 1986, pp. 304-324.

[14] G. E. Lucas, G. R. Odette, J. W. Sheckherd, and M. R. Krishnadev, "Recent Progress in Subsized Charpy Impact Specimen Testing for Fusion Reactor Materials Development," *Fusion Technology* Vol. 10, 1986, pp. 728-733.

[15] B. S. Louden, A. S. Kumar, F. A. Garner, M. L. Hamilton, and W. L. Hu, "The influence of specimen size on charpy impact testing of unirradiated HT-9," *Journal of Nuclear Materials* Vol. 155-157, 1988, pp. 662-667.

[16] A. Fabry, *et al.*, "Research to Understand the Embrittlement Behavior of Yankee/BR3 Surveillance Plate and Other Outlier RPV Steels," in *Effects of Radiation on Materials: 17th International Symposium*, ASTM STP 1270, D. S. Gelles, R. K. Nanstad, A. E. Kumar, and E. A. Liltle, Eds., ASTM, West Conshohocken, PA, 1996, pp. 138-190.

[17] R. C. Paule and J. Mandel, "Consensus values and weighting factors," *NIST Journal of Research*, 87 (5), 1982, pp. 303-308.

Annex A

Complete Round-Robin test results

Lab #1

Energy level	Specimen id	F_{gy} (kN)	F_m (kN)	F_{iu} (kN)	F_a (kN)	W_{gy} (J)	W_m (J)	W_{iu} (J)	W_a (J)	W_t (J)	KV (J)	KV/W_t	$KV-W_t$ (J)	LE (mm)	SFA (%)	SFA_{est} (%)
Low	LL8	1.41	2.32	2.28	1.18	0.08	0.80	0.96	0 98	1 32	1.55	1.174	0.23		39.1	60%
	LL23	1.58	2.35	2.33	1.15	0.10	0.85	0.97	1 02	1.36	1.57	1.154	0.21		43.4	57%
	LL26	1.58	2.34	2.29	1.14	0.11	0.80	0.96	0 99	1.31	1.51	1.153	0.20		43.1	58%
	LL31	1.48	2.34	2.33	1.04	0.10	0.87	0 96	1 00	1.33	1.52	1.143	0.19		42.3	53%
	LL35	1.52	2.32	2.25	1.16	0.10	0.74	0 93	0 97	1.28	1.46	1.141	0.18		41.4	60%
Average values		1.51	2.33	2.30	1.13	0.10	0.81	0.96	0.99	1.32	1.52	1.153	0.20		41.9	58%
Standard deviations		0 072	0.013	0.034	0.055	0.011	0.051	0.015	0.019	0.029	0 042	0 013	0.02		1.73	0.028
High	HH8	1.39	1.75			0.10	1 35			4.85	5.53	1.140	0.68			
	HH24	1.37	1.74			0.09	1.43			4.85	5.57	1.148	0.72			
	HH28	1.40	1.76			0.10	1.40			4.92	5.64	1.146	0.72			
	HH36	1.35	1.72			0.10	1 36			4.81	5.53	1.150	0.72			
	HH38	1.36	1.72			0.10	1 39			4.95	5.65	1.141	0.70			
Average values		1.37	1.74			0.10	1.39			4.88	5.58	1.145	0.71			
Standard deviations		0.021	0.018			0.004	0.032			0 057	0 058	0.004	0.02			
Super High	SH3	1.14	1.77			0.08	1 04			8.83	10.15	1.149	1.32			
	SH7	1.18	1.74			0.08	1 38			8.91	10.16	1.140	1.25			
	SH11	1.21	1.77			0.10	1 32			9.17	10.44	1.138	1.27			
	SH28	1.19	1.79			0.10	1 35			8.95	10.29	1.150	1.34			
	SH34	1.14	1.76			0 08	1 31			9.03	10.26	1.136	1.23			
Average values		1.17	1.77			0.09	1.28			8.98	10.26	1.143	1.28			
Standard deviations		0.031	0.018			0.011	0.137			0.129	0.118	0.006	0.047			

Lab #2

Energy level	Specimen id	F_{gy} (kN)	F_m (kN)	F_{iu} (kN)	F_a (kN)	W_{gy} (J)	W_m (J)	W_{iu} (J)	W_a (J)	W_t (J)	KV (J)	KV/W_t	$KV-W_t$ (J)	LE (mm)	SFA_{meas} (%)	SFA_{est} (%)
Low	LL3	1.95	2.37	2.32	1.33	0.33	0.70	0.96	0 99	1.48	1.60	1.082	0.12			62%
	LL9	1.99	2.35	2.32	1.34	0.27	0.73	0.88	0 94	1.37	1.49	1 089	0.12			61%
	LL15	1.95	2.37	2.33	1.46	0.27	0.66	0 89	0 96	1.41	1.51	1 069	0.10			66%
	LL18	1.57	2.38	2.29	1.24	0.20	0.78	0 98	1 02	1.46	1.58	1 081	0.12			62%
	LL28	1.87	2.31	2.17	1.32	0.27	0.64	0 91	0 97	1.44	1.55	1 077	0.11			66%
Average values		**1.87**	**2.35**	**2.29**	**1.34**	**0.27**	**0.70**	**0.92**	**0.98**	**1.43**	**1.55**	**1.080**	**0.11**			**64%**
Standard deviations		0.171	0.028	0.065	0.080	0.045	0.056	0.043	0.029	0.043	0 046	0 007	0.010			0.026
High	HH4	1.50	1.83			0.33	1 56			5.84	5.88	1 007	0.04			
	HH7	1.63	1.78			0.44	1.68			5.87	5.86	0.998	-0 01			
	HH10	1.40	1.77			0.31	1 39			5.46	5.57	1.021	0.11			
	HH12	1.46	1.80			0.32	1.41			5.43	5.47	1.008	0.04			
	HH18	1.49	1.80			0.29	1.66			5.62	5.72	1.017	0.10			
Average values		**1.50**	**1.79**			**0.34**	**1.54**			**5.64**	**5.70**	**1.010**	**0.06**			
Standard deviations		0.087	0.025			0.057	0.135			0 208	0.179	0.009	0.050			
Super High	SH13	1.40	1.78			0 20	1.42			9.88	9.97	1.009	0.09			
	SH14	1.52	1.78			0 27	1.63			9.99	10.02	1.003	0.03			
	SH18	1.41	1.81			0.19	1.62			9.48	9.48	1.000	0.00			
	SH29	1.42	1.79			0 20	1 23			10.18	10.21	1.003	0.03			
	SH38	1.49	1.79			0 23	1.64			10.29	10.25	0.996	-0 04			
Average values		**1.45**	**1.79**			**0.22**	**1.51**			**9.96**	**9.99**	**1.002**	**0.02**			
Standard deviations		0.054	0.012			0.033	0.180			0 315	0.307	0.005	0.049			

Lab #3

Energy level	Specimen id	F_{gy} (kN)	F_m (kN)	F_{iu} (kN)	F_a (kN)	W_{gy} (J)	W_m (J)	W_{iu} (J)	W_a (J)	W_t (J)	KV (J)	KV/W_t	$KV-W_t$ (J)	LE (mm)	SFA$_{meas}$ (%)	SFA$_{est}$ (%)
Low	LL6	2.09	2.26	2.26	0.72	▪	0.96	0.96	▪	1 55	1.84	1.187	0.29	0.10	▪	34%
	LL14	2.14	2.17	2.17	0.66		0.94	0.94		1 37	1.72	1.255	0.35	0.09		31%
	LL25	2.16	2.17	2.17	0.32	▪	0.90	0.90	▪	1 33	1.72	1 293	0.39	0.03	▪	15%
	LL31	2.26	2.26	2.26	0.62	▪	0.95	0.95	▪	1.43	1.82	1 273	0.39	0.06	▪	27%
	LL39	2.04	2.29	2.29	0.53		1.00	1.00		1.45	1.79	1 234	0.34	0.05	▪	27%
Average values		**2.14**	**2.23**	**2.23**	**0.57**		**0.95**	**0.95**		**1.43**	**1.78**	**1.249**	**0.35**	**0.07**		**27%**
Standard deviations		0 083	0 056	0.056	0.157	▪	0.036	0.036	▪	0.084	0.056	0 041	0.04	0.03	▪	0.074
High	HH13	1.36	1.70			▪	1.47		▪	5.31	5.73	1 079	0.42	0.56		
	HH20	1.37	1.67	▪		▪	1.49	▪	▪	5.41	5.66	1 046	0.25	0.54		
	HH25	1.43	1.67	▪		▪	1.43	▪	▪	5.33	5.67	1 064	0.34	0.57		
	HH37	1.38	1.64	▪		▪	1.43	▪	▪	5.51	5.82	1 056	0.31	0.53		
	HH39	1.40	1.67	▪	▪	▪	1.44	▪	▪	5.51	5.86	1 064	0.35	0.58		
Average values		**1.39**	**1.67**				**1.45**			**5.41**	**5.75**	**1.062**	**0.33**	**0.56**		
Standard deviations		0 028	0.021	▪	▪	▪	0.027	▪	▪	0.095	0 089	0 012	0.06	0.02		
Super High	SH4	1.51	1.68	▪	▪	▪	1.77	▪	▪	9.85	10.00	1 015	0.15	1.01		
	SH6	1.55	1.70				1.77			9.69	10.17	1.050	0.48	0.97		
	SH15	1.52	1.69	▪	▪	▪	1 85	▪	▪	9.98	10.42	1 044	0.44	1.03		
	SH32	1.49	1.69	▪	▪	▪	1.79	▪	▪	9.72	10.17	1 046	0.45	0.96		
	SH33	1.54	1.70		▪	▪	1 89	▪	▪	9.79	10.00	1 021	0.21	1.02		
Average values		**1.52**	**1.69**	▪	▪	▪	**1.81**	▪	▪	**9.81**	**10.15**	**1.035**	**0.35**	**1.00**		
Standard deviations		0.022	0.009				0.054			0.115	0.172	0.016	0.154	0.031		

Lab #4

Energy level	Specimen id	F_{gy} (kN)	F_m (kN)	F_{lu} (kN)	F_a (kN)	W_{gy} (J)	W_m (J)	W_{lu} (J)	W_a (J)	W_t (J)	KV (J)	KV/W_t	$KV-W_t$ (J)	LE (mm)	SFA_{meas} (%)	SFA_{est} (%)
Low	LL2	2.01	2.83	2.47	0.52	0.31	0.83	1.01	1 22	1 57	1.62	1 032	0.05			40%
	LL12	1.94	2.79	2.57	0.82	0.29	0.83	0.95	1.15	1.44	1.50	1 042	0.06			46%
	LL20	1.95	2.75	2.54	0.63	0.28	0.82	0.94	1 22	1.50	1.58	1 053	0.08			39%
	LL22	1.96	2.80	2.51	0.46	0.30	0.82	0 94	1.15	1.54	1.58	1 026	0.04			36%
	LL40	2.03	2.85	2.56	0.14	0.31	0.84	1 02	1 31	1.56	1.59	1 019	0.03			26%
Average values		**1.98**	**2.80**	**2.53**	**0.51**	**0.30**	**0.83**	**0.97**	**1.21**	**1.52**	**1.57**	**1.034**	**0.05**			**37%**
Standard deviations		0 038	0.039	0.039	0.250	0.013	0.008	0.040	0.066	0.053	0 044	0 013	0.02			0.073
High	HH5	1.63	1.90			0.22	1.67			5.93	5.79	0 976	-0.14			
	HH9	1.63	1.92			0.22	1.45			5.76	5.62	0 976	-0.14			
	HH15	1.56	1.89			0.18	1.44			5.64	5.43	0 963	-0.21			
	HH16	1.61	1.93			0.20	1.64			5.89	5.75	0 976	-0.14			
	HH21	1.61	1.88			0.19	1.61			5.79	5.58	0.964	-0 21			
Average values		**1.61**	**1.90**			**0.20**	**1.56**			**5.80**	**5.63**	**0.971**	**-0.17**			
Standard deviations		0.029	0.022			0.018	0.109			0.114	0.144	0.007	0.04			
Super High	SH21	1.71	1.93			0.27	1 59			10.59	10.13	0.957	-0.46			
	SH23	1.69	1.90			0.24	1.64			10.64	10.05	0.945	-0 59			
	SH27	1.69	1.89			0.25	1.61			10.14	9.60	0.947	-0 54			
	SH30	1.70	1.91			0.25	1.69			10.29	9.73	0.946	-0 56			
	SH37	1.69	1.90			0.24	1 58			10.32	9.79	0.949	-0 53			
Average values		**1.70**	**1.90**			**0.25**	**1.62**			**10.40**	**9.86**	**0.948**	**-0.54**			
Standard deviations		0.009	0.014			0.012	0.044			0 212	0 223	0.005	0.048			

Lab #5

Energy level	Specimen id	F_{gy} (kN)	F_m (kN)	F_{lu} (kN)	F_a (kN)	W_{gy} (J)	W_m (J)	W_{lu} (J)	W_a (J)	W_t (J)	KV (J)	KV/W_t	$KV-W_t$ (J)	LE (mm)	SFA$_{meas}$ (%)	SFA$_{est}$ (%)
Low	LL47	1.86	2.69	2.62	1.67	0.28	1.12	1.12	1.17	1.44	1.35	0.938	-0.09	0.07		69%
	LL50	1.93	2.72	2.66	1.73	0.32	1.14	1.14	1 22	1.45	1.39	0 959	-0.06	0.06		70%
	LL62	1.93	2.69	2.62	1.71	0.32	1.12	1.22	1 26	1.57	1.52	0 968	-0.05	0.07		70%
	LL65	1.74	2.73	2.71	1.70	0.25	1.17	1.17	1 24	1.50				0.04		69%
	LL80	2.18	2.70	2.65	1.46	0.42	1.11	1.11	1 20	1.43	1.34	0 937	-0.09	0.05		60%
Average values		**1.93**	**2.71**	**2.65**	**1.65**	**0.32**	**1.13**	**1.15**	**1.22**	**1.48**	**1.40**	**0.950**	**-0.07**	**0.06**		**68%**
Standard deviations		0.161	0.018	0.037	0.111	0.064	0.024	0.044	0.035	0.058	0 083	0 016	0.021	0.013		0.045
High	HH45	1.57	1.91			0.22	2.69			6.43	5.52	0 858	-0.91	0.51		
	HH54	1.68	1.98			0.28	2.17			6.27	5.38	0 858	-0.89	0.46		
	HH64	1.63	1.89			0.27	1.47			6.30	5.45	0 865	-0.85	0.47		
	HH72	1.48	1.89			0.23	1 32			6.20	5.40	0 871	-0.80	0.47		
	HH76	1.68	1.98			0.27	2.16			6.45	5.57	0 864	-0.88	0.52		
Average values		**1.61**	**1.93**			**0.25**	**1.96**			**6.33**	**5.46**	**0.863**	**-0.87**	**0.49**		
Standard deviations		0 085	0.046			0.027	0.563			0.107	0 080	0 005	0.043	0.027		
Super High	SH45	1.67	1.95			0.30	1 83			11 53	9.96	0.864	-1 57	0.95		
	SH56	1.67	1.97			0.32	2 25			10.77	9.62	0.893	-1.15	0.88		
	SH62	1.65	1.98			0.29	2 25			11.29	10.07	0.892	-1 22	0.90		
	SH69	1.66	1.97			0.29	2 24			10.91	9.65	0.885	-1 26	0.91		
	SH80	1.65	1.97			0.29	2 23			11.06	9.87	0.892	-1.19	0.95		
Average values		**1.66**	**1.97**			**0.30**	**2.16**			**11.11**	**9.83**	**0.885**	**-1.28**	**0.92**		
Standard deviations		0.010	0.011			0.013	0.185			0 303	0.195	0.012	0.168	0.031		

Lab #6

Energy level	Specimen id	F_{gy} (kN)	F_m (kN)	F_{iu} (kN)	F_a (kN)	W_{gy} (J)	W_m (J)	W_{iu} (J)	W_a (J)	W_t (J)	KV (J)	KV/W_t	$KV-W_t$ (J)	LE (mm)	SFA_{meas} (%)	SFA_{est} (%)
Low	LL43	1.58	2.27	2.20	0.93	0.22	0.77	0.93	0 96	1.45	1.80	1.241	0.35	0.07	40	51%
	LL55	1.54	2.24	2.16	0.71	0.20	0.75	0.92	0 95	1.43	2.00	1 399	0.57	0.04	40	44%
	LL60	1.55	2.26	2.16	1.00	0.21	0.79	1 01	1 07	1.64	2.00	1 220	0.36	0.06	50	56%
	LL72	1.44	2.25	2.20	0.77	0.17	0.80	0 96	0 98	1.48	1.90	1 284	0.42	0.09	50	46%
	LL73	1.41	2.28	2.22	0.80	0.16	0.80	0 97	0 99	1.43	1.80	1 259	0.37	0.05	50	48%
Average values		**1.50**	**2.26**	**2.19**	**0.84**	**0.19**	**0.78**	**0.96**	**0.99**	**1.49**	**1.90**	**1.280**	**0.41**	**0.06**	**46**	**49%**
Standard deviations		0 074	0.016	0.027	0.119	0.026	0.022	0.036	0.047	0.088	0.100	0 070	0.091	0.019	5.477	0.046
High	HH43	1.32	1.70			0.20	1.44			5.49	5.90	1.075	0.41	0.46	70	
	HH53	1.36	1.71			0.23	1.46			5.43	5.80	1.068	0.37	0.35	70	
	HH60	1.37	1.70			0.23	1.40			5.57	6.00	1.077	0.43	0.54	70	
	HH71	1.39	1.71			0.26	1.46			5.48	5.90	1.077	0.42	0.37	80	
	HH74	1.36	1.70			0.22	1.42			5.58	6.00	1.075	0.42	0.55	70	
Average values		**1.36**	**1.70**			**0.23**	**1.44**			**5.51**	**5.92**	**1.074**	**0.41**	**0.45**	**72**	
Standard deviations		0.025	0.005			0.022	0.026			0 064	0 084	0.004	0.023	0.093	4.472	
Super High	SH48	1.41	1.71			0 24	2 02			9.59	10.00	1.043	0.41	0.75	95	
	SH54	1.42	1.71			0 24	2 03			9.76	10.10	1.035	0.34	0.71	95	
	SH57	1.43	1.70			0 23	2.10			10.25	10.70	1.044	0.45	0.85	95	
	SH67	1.42	1.68			0 23	2.09			10.05	10.50	1.045	0.45	0.65	95	
	SH78	1.42	1.69			0 24	2.10			10.09	10.50	1.041	0.41	0.72	95	
Average values		**1.42**	**1.70**			**0.24**	**2.07**			**9.95**	**10.36**	**1.041**	**0.41**	**0.74**	**95**	
Standard deviations		0.007	0.013			0.005	0.040			0.267	0.297	0.004	0.045	0.073	0.000	

Lab #7

Energy level	Specimen id	F_{gy} (kN)	F_m (kN)	F_{iu} (kN)	F_a (kN)	W_{gy} (J)	W_m (J)	W_{iu} (J)	W_a (J)	W_t (J)	KV (J)	KV/W_t	$KV-W_t$ (J)	LE (mm)	SFA_{meas} (%)	SFA_{est} (%)
Low	LL5	1.77	2.34	2.27	1.52	0.26	0.72	0.89	0 99	1.46	1.55	1.057	0.08			72%
	LL10	1.67	2.31	2.26	1.43	0.24	0.74	0.90	1 01	1.46	1.53	1 042	0.06			68%
	LL16	1.80	2.40	2.30	1.45	0.27	0.75	0 87	0 96	1.37	1.42	1 037	0.05			69%
	LL30	1.77	2.35	2.23	1.45	0.26	0.64	0 88	0 97	1.37	1.43	1 039	0.05			70%
	LL37	1.79	2.33	2.21	1.53	0.26	0.73	0 85	0 96	1.42	1.49	1 049	0.07			74%
Average values		1.76	2.35	2.25	1.48	0.26	0.72	0.88	0.98	1.42	1.48	1.045	0.06			71%
Standard deviations		0 051	0.036	0.032	0.047	0.010	0.044	0.018	0.021	0.047	0 058	0 008	0.01			0.022
High	HH3	1.39	1.67			0.19	1 21			5.74	5.53	0 963	-0 21			
	HH26	1.37	1.66			0.17	1 35			5.34	5.43	1.018	0.09			
	HH31	1.31	1.69			0.15	1.12			5.37	5.51	1.026	0.14			
	HH32	1.33	1.65			0.11	1 24			5.43	5.54	1.020	0.11			
	HH34	1.37	1.66			0.14	1 26			5.33	5.46	1.025	0.13			
Average values		1.35	1.67			0.15	1.24			5.44	5.49	1.010	0.05			
Standard deviations		0.033	0.015			0.032	0.086			0.172	0 045	0.027	0.15			
Super High	SH8	1.45	1.64			0.26	1.76			9.42	9.62	1.021	0.20			
	SH11	1.42	1.69			0 23	1 95			9.90	10.01	1.010	0.10			
	SH20	1.44	1.68			0 24	1 85			9.93	10.01	1.008	0.07			
	SH26	1.43	1.69			0 23	1 20			9.43	9.60	1.018	0.17			
	SH40	1.42	1.66			0 31	1 87			9.92	9.86	0.994	-0 06			
Average values		1.43	1.67			0.25	1.73			9.72	9.82	1.010	0.10			
Standard deviations		0.014	0.020			0.030	0.301			0 270	0.199	0.011	0.103			

Lab #8

Energy level	Specimen id	F_{gy} (kN)	F_m (kN)	F_{lu} (kN)	F_a (kN)	W_{gy} (J)	W_m (J)	W_{lu} (J)	W_a (J)	W_t (J)	KV (J)	KV/W_t	$KV-W_t$ (J)	LE (mm)	SFA_{meas} (%)	SFA_{est} (%)
Low	LL4	2.04	2.45	2.45	1.64		0.98	0.98	1 04	1.67	1.63	0.980	-0.03			70%
	LL7	2.00	2.46	2.46	1.46		0.98	0.98	1 05	1.65	1.61	0 979	-0.04			63%
	LL11	1.72	2.36	2.36	1.62		0.98	0.98	1 01	1 58	1.55	0 981	-0.03			72%
	LL17	2.04	2.35	2.35	1.33		0.98	0.98	1 03	1.58	1.54	0 978	-0.03			59%
	LL24	2.01	2.47	2.47	1.69		0.99	0.99	1 09	1.72	1.68	0 980	-0.03			71%
Average values		**1.96**	**2.42**	**2.42**	**1.55**		**0.98**	**0.98**	**1.05**	**1.64**	**1.60**	**0.979**	**-0.03**			**67%**
Standard deviations		0.135	0 057	0.057	0.149		0.004	0.004	0.030	0.060	0.059	0 001	0.00			0.057
High	HH2	1.25	1.80				1.68			5.92	5.78	0 976	-0.14			
	HH19	1.38	1.77				1.49			5.74	5.61	0 978	-0.13			
	HH23	1.52	1.78				1.66			5.71	5.59	0 979	-0.12			
	HH35	1.25	1.74				1.45			5.58	5.46	0 978	-0.12			
	HH40	1.32	1.75				1.57			5.76	5.64	0 978	-0.13			
Average values		**1.34**	**1.77**				**1.57**			**5.74**	**5.61**	**0.978**	**-0.13**			
Standard deviations		0.110	0.025				0.100			0.121	0.115	0 001	0.01			
Super High	SH9	1.21	1.79				1.91			10 03	9.87	0 984	-0.16			
	SH12	1.26	1.79				2.08			9.98	9.86	0 987	-0.13			
	SH19	1.41	1.80				2 00			10.47	10.28	0 982	-0.19			
	SH24	1.28	1.79				1 98			10.19	10.06	0 987	-0.13			
	SH25	1.31	1.79				1 90			10.40	10.27	0 987	-0.14			
Average values		**1.29**	**1.79**				**1.98**			**10.22**	**10.07**	**0.985**	**-0.15**			
Standard deviations		0 074	0.007				0.075			0.219	0 207	0 003	0.028			

Lab #9

Energy level	Specimen id	F_{gy} (kN)	F_m (kN)	F_{lu} (kN)	F_a (kN)	W_{gy} (J)	W_m (J)	W_{iu} (J)	W_a (J)	W_t (J)	KV (J)	KV/W_t	$KV-W_t$ (J)	LE (mm)	SFA (%)	SFA_{est} (%)
Low	LL1		2.43	2.43	1.61		0.87	0.87	1 00	1 53	1.56	1.020	0.03	0.11	44.6	66%
	LL21		2.45	2.45	1.45		0.86	0.86	1.10	1.47	1.52	1 034	0.05	0.10	46.2	59%
	LL33		2.39	2.39	1.56		0.88	0.88	0 97	1.40	1.43	1 021	0.03	0.09	52.3	65%
	LL36		2.44	2.44	1.37		0.86	0 86	1.10	1.53	1.54	1 007	0.01	0.09	58.8	56%
	LL38		2.43	2.43	1.53		0.91	0 91	1.10	1.51	1.54	1 020	0.03	0.10	38.8	63%
Average values			**2.43**	**2.43**	**1.50**		**0.88**	**0.88**	**1.05**	**1.49**	**1.52**	**1.020**	**0.03**	**0.10**	**48.1**	**62%**
Standard deviations			0.023	0.023	0.095		0.021	0.021	0.064	0.055	0 051	0 010	0.01	0.008	7.66	0.042
High	HH 6	1.36	1.85			0.13	1.60			5.99	5.81	0 970	-0.18	0.60		
	HH17	1.37	1.84			0.13	1.40			5.67	5.57	0 982	-0.10	0.56		
	HH22	1.38	1.84			0.13	1 50			5.75	5.59	0 972	-0.16	0.55		
	HH27	1.37	1.83			0.13	1.40			5.74	5.62	0 979	-0.12	0.57		
	HH30	1.40	1.84			0.17	1 50			5.68	5.49	0 967	-0.19	0.54		
Average values		**1.38**	**1.84**			**0.14**	**1.48**			**5.77**	**5.62**	**0.974**	**-0.15**	**0.56**		
Standard deviations		0 015	0.007			0.018	0.084			0.130	0.119	0 007	0.04	0.023		
Super High	SH1	1.44	1.84			0.24	1.60			10.43	10.11	0.969	-0 32	1.02		
	SH22	1.45	1.84			0.23	1 50			10.62	10.26	0.966	-0 36	1.04		
	SH31	1.40	1.85			0.23	1 50			10.12	9.86	0.974	-0 26	1.02		
	SH35	1.45	1.86			0.24	1 50			10.20	9.88	0.969	-0 32	1.05		
	SH39	1.44	1.85			0.24	1 50			10.82	10.46	0.967	-0 36	1.06		
Average values		**1.44**	**1.85**			**0.24**	**1.52**			**10.44**	**10.11**	**0.969**	**-0.32**	**1.04**		
Standard deviations		0.021	0.008			0.005	0.045			0 290	0 255	0.003	0.041	0.018		

Annex B

Detailed calculations
for the establishment of
certified reference values

Low Energy specimens - KV (J)

Lab	n_i	$\bar{Y}_i$	s_i^2	s_i^2/n_i	ω_i	$(\bar{Y}_i - \hat{Y})^2$	$s_i^2/n_i + s_b^2$	Ratio	$\omega_i\bar{Y}_i$
1	5	1.52	0.00177	0.000354	43.867515	0.004657	0.022796	0.204273	66.766357
2	5	1.55	0.00213	0.000426	43.729397	0.001957	0.022868	0.085583	67.605648
3	5	1.78	0.00312	0.000624	43.354020	0.035254	0.023066	1.528408	77.083447
4	5	1.57	0.00198	0.000396	43.786840	0.000264	0.022838	0.011547	68.920486
5	4	1.40	0.00687	0.001717	41.393164	0.036191	0.024159	1.498057	57.950429
6	5	1.90	0.01000	0.002000	40.913330	0.095952	0.024442	3.925707	77.735326
7	5	1.48	0.00336	0.000672	43.264473	0.011837	0.023114	0.512133	64.093721
8	5	1.60	0.00344	0.000688	43.233536	0.000163	0.023130	0.007068	69.304456
9	5	1.52	0.00262	0.000524	43.542795	0.005218	0.022966	0.227228	66.097963

Between-lab variance $s_b^2 = 0.0224$

Consensus mean $\hat{Y} =$	1.5902 J
Mean of the means $\bar{Y} =$	1.5914 J
Average all results $\hat{Y} =$	1.5957 J

Stand. uncertainty $u(\hat{Y}) = 0.0508$ J

Low Energy specimens - F_m (kN)

Lab	n_i	$\bar{Y}_i$	s_i^2	s_i^2/n_i	ω_i	$(\bar{Y}_i - \hat{Y})^2$	$s_i^2/n_i + s_b^2$	Ratio	$\omega_i\bar{Y}_i$
1	5	2.33	0.00018	0.000036	26.308187	0.009429	0.038011	0.248059	61.403307
2	5	2.35	0.00078	0.000156	26.225146	0.006038	0.038131	0.158341	61.718258
3	5	2.23	0.00312	0.000624	25.907218	0.039642	0.038599	1.027013	57.824910
4	5	2.80	0.00155	0.000310	26.120039	0.138307	0.038285	3.612596	73.214470
5	5	2.71	0.00033	0.000066	26.287439	0.075568	0.038041	1.986500	71.133811
6	5	2.26	0.00025	0.000050	26.298500	0.029276	0.038025	0.769920	59.434611
7	5	2.35	0.00127	0.000255	26.157640	0.007345	0.03823	0.192128	61.350128
8	5	2.42	0.00325	0.000651	25.889488	0.000201	0.038626	0.005194	62.573297
9	5	2.43	0.00052	0.000104	26.261206	0.000010	0.038079	0.000253	63.762209

Between-lab variance $s_b^2 = 0.0380$

Consensus mean $\hat{Y} =$	2.4311 kN
Mean of the means $\bar{Y} =$	2.4310 kN
Average all results $\hat{Y} =$	2.4310 kN

Stand. uncertainty $u(\hat{Y}) = 0.0652$ kN

High Energy specimens - KV (J)

Lab	n_i	$\bar{Y}_i$	s_i^2	s_i^2/n_i	ω_i	$(\bar{Y}_i - \hat{Y})^2$	$s_i^2/n_i + s_b^2$	Ratio	$\omega_i\bar{Y}_i$
1	5	5.58	0.00338	0.000676	67.540345	0.004499	0.014806	0.303859	377.145286
2	5	5.70	0.03205	0.006410	48.685574	0.002394	0.020540	0.116541	277.507772
3	5	5.75	0.00797	0.001594	63.597190	0.009395	0.015724	0.597473	365.556646
4	5	5.63	0.02063	0.004126	54.776616	0.000292	0.018256	0.015969	308.611455
5	5	5.57	0.00643	0.001286	64.867816	0.006573	0.015416	0.426376	361.313736
6	5	5.92	0.00700	0.001400	64.391644	0.072321	0.015530	4.656880	381.198535
7	5	5.49	0.00206	0.000413	68.763789	0.025196	0.014543	1.732606	377.674107
8	5	5.61	0.01314	0.002628	59.673620	0.001302	0.016758	0.077704	335.066709
9	5	5.62	0.01408	0.002816	59.011097	0.00123	0.016946	0.072595	331.406321

Between-lab variance $s_b^2 = 0.0141$

Consensus mean $\hat{Y} =$	5.6511 J
Mean of the means $\bar{Y} =$	5.6533 J
Average all results $\hat{Y} =$	5.6415 J

Stand. uncertainty $u(\hat{Y}) = 0.0426$ J

High Energy specimens - F_m (kN)

Lab	n_i	$\bar{Y}_i$	s_i^2	s_i^2/n_i	ω_i	$(\bar{Y}_i - \hat{Y})^2$	$s_i^2/n_i + s_b^2$	Ratio	$\omega_i\bar{Y}_i$
1	5	1.74	0.00032	0.000064	90.076780	0.002142	0.011102	0.192954	156.553444
2	5	1.79	0.00063	0.000126	89.574431	0.000087	0.011164	0.007776	160.660699
3	5	1.67	0.00043	0.000087	89.894265	0.012878	0.011124	1.157691	150.195338
4	5	1.90	0.00048	0.000096	89.821434	0.013530	0.011133	1.215254	170.714617
5	5	1.98	0.00215	0.000430	87.201899	0.038305	0.011468	3.340285	172.659760
6	5	1.70	0.00003	0.000006	90.549854	0.006445	0.011044	0.583625	154.296951
7	5	1.67	0.00022	0.000044	90.238006	0.013313	0.011082	1.201358	150.598208
8	5	1.77	0.00063	0.000125	89.581293	0.000224	0.011163	0.020062	158.497776
9	5	1.84	0.00005	0.000010	90.517068	0.003104	0.011048	0.281001	166.551406

Between-lab variance $s_b^2 = 0.0110$

Consensus mean $\hat{Y} =$	1.7843 kN
Mean of the means $\bar{Y} =$	1.7850 kN
Average all results $\hat{Y} =$	1.7795 kN

Stand. uncertainty $u(\hat{Y}) = 0.0352$ kN

Super-High Energy specimens - KV (J)

Lab	n_i	$\bar{Y}_i$	s_i^2	s_i^2/n_i	ω_i	$(\bar{Y}_i - \hat{Y})^2$	$s_i^2/n_i + s_b^2$	Ratio	$\omega_i\bar{Y}_i$
1	5	10.26	0.01385	0.002770	22.774796	0.053976	0.043908	1.229297	233.669403
2	5	9.99	0.09433	0.018866	16.665503	0.001737	0.060004	0.028941	166.421718
3	5	10.15	0.02967	0.005934	21.243967	0.015457	0.047072	0.328376	215.668752
4	5	9.86	0.04960	0.009920	19.585497	0.028114	0.051058	0.550626	193.113004
5	5	9.65	0.03813	0.007626	20.506852	0.142636	0.048764	2.925022	197.891126
6	5	10.36	0.08800	0.017600	17.024700	0.110442	0.058738	1.880237	176.375888
7	5	9.82	0.03950	0.007901	20.391881	0.043569	0.049039	0.888457	200.226659
8	5	10.07	0.04268	0.008535	20.131488	0.001561	0.049673	0.031428	202.667381
9	5	10.11	0.06508	0.013016	18.465793	0.007452	0.054154	0.137616	186.763027

Between-lab variance $s_b^2 = 0.0411$

Consensus mean $\hat{Y} =$	10.0277 J
Mean of the means $\bar{Y} =$	10.0298 J
Average all results $\hat{Y} =$	10.0502 J

Stand. uncertainty $u(\hat{Y}) = 0.0752$ J

Super-High Energy specimens - F_m (kN)

Lab	n_i	$\bar{Y}_i$	s_i^2	s_i^2/n_i	ω_i	$(\bar{Y}_i - \hat{Y})^2$	$s_i^2/n_i + s_b^2$	Ratio	$\omega_i\bar{Y}_i$
1	5	1.77	0.00033	0.000066	98.228059	0.000694	0.010180	0.068140	173.470752
2	5	1.79	0.00015	0.000030	98.576647	0.000005	0.010144	0.000539	176.452198
3	5	1.69	0.00008	0.000017	98.703719	0.010229	0.010131	1.009629	166.927730
4	5	1.90	0.00021	0.000042	98.463666	0.012513	0.010156	1.232088	187.494514
5	5	1.97	0.00012	0.000024	98.634985	0.031564	0.010138	3.113296	194.310921
6	5	1.70	0.00017	0.000034	98.537793	0.008900	0.010148	0.876952	167.317172
7	5	1.67	0.00042	0.000083	98.063096	0.014203	0.010198	1.392828	164.075249
8	5	1.79	0.00004	0.000008	98.786501	0.000006	0.010123	0.000636	176.808198
9	5	1.85	0.00007	0.000014	98.732370	0.003098	0.010128	0.305899	182.457420

Between-lab variance $s_b^2 = 0.0101$

Consensus mean $\hat{Y} =$	1.7923 kN
Mean of the means $\bar{Y} =$	1.7923 kN
Average all results $\hat{Y} =$	1.7920 kN

Stand. uncertainty $u(\hat{Y}) = 0.0336$ kN